bibliographia geodaetica

AKADEMIE-VERLAG · BERLIN

Volume 22 · 1984 · Number 2

ISSN 0006–1239 bibl. geod., Berlin **22** (1984) 2, 97–188 EVP 15,– M

CONTENTS

BIBLIOGRAPHIA GEODAETICA

International Geodetic Documentation

Booklet of reports of the
International Association of Geodesy
International Federation of Surveyors
Technical University of Dresden, GDR

Published by the National Committee of Geodesy and Geophysics of the Academy of Sciences of the German Democratic Republic

Prof. Dr.-Ing. Horst Peschel, president

Edition by the Centre of International Documentation of Geodesy, Technical University of Dresden, Section Geodesy and Cartography

Dr.-Ing. Heinz Finger, manager

AKADEMIE-VERLAG · BERLIN

bibliographia geodaetica	Volume 22 1984 No. 2	p. 97–188	Berlin 1984

Sources of supplies

Orders and subscriptions of the "bibliographia geodaetica" should be sent:

- in the German Democratic Republic to a book-shop, or to the Akademie-Verlag, DDR – 1086 Berlin, Leipziger Str. 3–4;
- in socialist countries to a book-shop for foreign-language literature, or to the responsible post-office;
- in the Federal Republic of Germany and Berlin (West) to a book-shop, or to the supply depot KUNST UND WISSEN, Erich Bieber OHG, Wilhelmstr. 4–6, D – 7000 Stuttgart 1;
- in other West-European countries to a book-shop, or to the supply depot KUNST UND WISSEN, Erich Bieber GmbH, Dufourstr. 51, CH – 8008 Zürich;
- in the remaining foreign countries to the International Book and Journal Trade; to the Buchexport, Volkseigener Aussenhandelsbetrieb der Deutschen Demokratischen Republik, DDR – 7010 Leipzig, Postfach 160, or to the Akademie-Verlag, DDR – 1086 Berlin, Leipziger Str. 3–4.

Information of additional sources of supplies is given by the Bureau Central de l'AIG, 39ter, rue Gay Lussac, F – 75005 Paris.

bibliographia geodaetica

Herausgeber: Nationalkomitee für Geodäsie und Geophysik der Akademie der Wissenschaften der Deutschen Demokratischen Republik.
Verlag: Akademie-Verlag, DDR – 1086 Berlin, Leipziger Straße 3–4; Fernruf: 2 23 62 21 oder 2 23 62 29; Telex Nr.: 11 44 20; Bank: Staatsbank der DDR, Berlin, Kto.-Nr.: 6836-26-20712.
Chefredakteur: Dr.-Ing. Heinz Finger.
Redaktion: Zentralstelle für Internationale Dokumentation der Geodäsie, Technische Universität Dresden, Sektion Geodäsie und Kartographie, DDR – 8027 Dresden, Mommsenstraße 13; Fernruf: 4 63 28 23 (RIS 1107).
Druckgenehmigung: Ag 451/47/84.
Satz und Druck: Druckwerkstätten Stollberg VOB, DDR – 9150 Stollberg (Erzgeb.) - III/6/50 484 (413).
Erscheinungsweise: Die „bibliographia geodaetica" erscheint jeden zweiten Monat. Bezugspreis eines Jahrganges 120,– M zuzüglich Versandspesen (Preis für die DDR 90,– M).
Bestellnummer dieses Heftes: 1073/22/2.

BIBLIOGRAPHIA GEODAETICA

Authors

(The figures behind the authors indicate the number of the abstract; the Roman numerals indicate the group of the abstract)

No. 2/1984

bibliographia geodaetica

New series, Volume 22, No. 2 1984

AIDOS Systematics **12.01** **GENERALITIES**

General problems of geodesy, surveying, photogrammetry and cartography

Management, steering, cooperation with other branches of economy
History of development
National scientific-technical organizations
International co-operation
Research, invention affairs, innovators
Information activities, reference books, terminology
Standardization, metrological ensuring
Organization, planning, economy
Staff and training
Quality proof and quality ensuring
Labour and health safety

ALLGEMEINES

Allgemeine Fragen der Geodäsie, Photogrammetrie und Kartographie

Leitung, Zusammenarbeit mit anderen Wirtschaftszweigen
Historische Entwicklung
Nationale wissenschaftlich-technische Organisationen
Internationale Zusammenarbeit
Forschung, Erfindungswesen, Neuerer
Informationstätigkeit, Nachschlagewerke, Terminologie
Standardisierung, metrologische Sicherstellung
Organisation, Planung, Ökonomie
Kader und Ausbildung
Qualitätskontrolle, -sicherung
Arbeits- und Gesundheitsschutz

528.3(435.2)(091):528.27

Augath, Wolfgang
Die moderne Grundlagenvermessung in Oldenburg und ihre Erneuerung
200 Jahre Oldenburger Landesvermessung / Festschrift zum 200jaehrigen Jubilaeum der Oldenburger Landesvermessung, Mai 1981 : Ed. Deutscher Verein fuer Vermessungswesen, Bezirksgruppe Oldenburg—Bremen. — Oldenburg (1981). — p. 126—144
Modern basic survey in Oldenburg and its renewal (GER)
AS 12.01.09/12.15.25
Key words: Geodetic control net, gravity net, Oldenburg

1) The geodetic control net in Oldenburg:
   Work of the Prussian geodetic survey.
   Renewal of the geodetic control net (1972).
2) The vertical control survey net in Oldenburg:
   Measuring epochs in the Oldenburg area.
   Development of measuring accuracy.
   Development of reference marks.
   Evidence of recent crustal movements.

3) The gravitational control survey net:
Gravity system of Potsdam and geophysical survey of the German Reich.
German gravity net.
Recent work and projects.

Schoen, National Correspondent, Federal Republic of Germany IGD 84/A 23

528(71)"1980"

Babbage, G.
Activities of the Geodetic Survey of Canada in 1980 (ENG)
Energy, Mines and Resources Canada, Surveys and Mapp. Branch. – Ottawa (1980). – p. 4–14 : 6 lit.
AS 12.01.21
Key words: Canada, geodetic survey

Techn. Univ. Dresden IGD 84/A 24

528(71)"1981"

Babbage, G.
Activities of the Geodetic Survey of Canada in 1981 (ENG)
Energy, Mines and Resources Canada, Surveys and Mapp. Branch. – Ottawa (1981). – p. 1–14 : 1 fig.
AS 12.01.21
Key words: Canada, geodetic survey

Techn. Univ. Dresden IGD 84/A 25

061.14.055:528(71)"1980"

. . .
Collected papers – geodetic survey (ENG)
Ottawa: Surveys and Mapp. Branch, Dep. of Energy, Mines and Resources. 1980. 114 p. : div. fig. div. tab. div. lit.
AS 12.01.13/12.15.17
Key words: Canada, geodesy

These papers have been prepared mainly by the staff of Geodetic Survey for conferences, technical and scientific publications and as internal reports. We hope that they will prove to be helpful and informative and will stimulate thinking and development in geodesy.

Contents:

Babbage, G. – Activities of the geodetic survey of Canada in 1980 (p. 1–14)

Boal, J. D. – Satellite Doppler surveying (p. 15–48)

Lachapelle, G.; Kouba, J. – Relationship between terrestrial and satellite Doppler systems (p. 49–64)

Penton, C. R. – Methods for vertical mapping control (p. 65–82)

Vamosi, S. – Calibration procedures for invar levelling rods in Canada (p. 83–97)

Wilkinson, G. A. – Report on the ISS evaluation project 1979 Saskatchewan (p. 98–114).

Techn. Univ. Dresden IGD 84/A 26

061.14.055:528(71)"1981"

. . .
Collected papers – geodetic survey (ENG)
Ottawa: Surveys and Mapp. Branch, Dep. of Energy, Mines and Resources. 1981. 231 p. : div. fig. div. tab. div. lit.
AS 12.01.13/12.15.37
Key words: Canada, geodesy

The principal aim of this edition of collected papers is to provide information regarding recent activities of the Geodetic Survey of Canada. Most of the papers result from research and studies on geodesy and higher surveying and have been published in scientific or technical journals; others stem from internal reports.

Contents:

Part 1

Babbage, G. — Activities of Geodetic Survey of Canada in 1981 (p. 1—14)

Babbage, G.; McLellan, C. D. — Status of the redefinition and readjustment of national horizontal and vertical geodetic networks and of the national geodetic data base (p. 15—37)

Beattie, D. S.; Henderson, J. P.; Pinch, M. C. — Evaluation, strengthening and adjustment of geodetic networks in Canada (p. 39—68)

Boal, J. D.; Vamosi, S. — Comparative evaluation of microprocessor controlled transit satellite Doppler receiving equipment (p. 69—91)

Lachapelle, G.; Mainville, A. — Least squares prediction of horizontal co-ordinate distortions in Canada (p. 93—114)

Techn. Univ. Dresden IGD 84/A 27

061.14.055:528(71)"1981"

. . .

Collected papers — geodetic survey (ENG)
Ottawa: Surveys and Mapp. Branch, Dep. of Energy, Mines and Resources. 1981. 231 p. : div. fig. div. tab. div. lit.
AS 12.01.13/12.15.09
Key words: Canada, geodesy

Contents:

Part 2

Lachapelle, G.; Mainville, A.; Schwarz, K. P. — Empirical investigations on the anisotropy of the free air gravity anomaly covariance function (p. 115—131)

Lachapelle, G.; McConnell, K. — On the vertical reference system of the Canadian national gravity data base (p. 133—148)

Steeves, R. R. — Statistical test for significance of peaks in the least squares spectrum (p. 149—165)

Steeves, R. R. — A note on the optimal design of geodetic networks (p. 167—173)

Steeves, R. R. — Estimation of gravity tilt response to atmospheric phenomena at the Frederiction tiltmetric station using a least squares response method (p. 175—180)

Webb, J. D.; Penney, R. C. — Six years of inertial surveying at geodetic survey of Canada (p. 181—199)

Wilkinson, G. A. — Report on Mekometer calibrations during the 1980 field season (p. 201—218)

Wilkinson, G. A.; Wong, R. V. C. — The use of azimuth control in inertial surveying to improve the accuracy on an L-shaped traverse (p. 219—231).

Techn. Univ. Dresden IGD 84/A 28

378.6(430-2.27)"47-01":528

. . .

Festschrift 100 Jahre Geodaetische Lehre und Forschung in Hannover
(Ed. Spellauge, Reinhard)
Wiss. Arb. Fachricht. Vermess.-Wesen Univ. Hannover. — Hannover (1981) 100. — 313 p. : div. fig. div. tab. div. lit.
Commemorative publication "100 years of geodetic science and research in Hanover" (GER)
AS 12.01.09
Key words: Geodesy, Technical University Hanover

The University of Hanover can look back this year on 150 years of history and is marking this event with an anniversary week in which the various faculties will present themselves to the public in talks and discussion groups. In addition to this week the Faculty of Surveying is holding a special anniversary looking back on 100 years of geodetic science and research in Hanover with this special publication to commemorate it.

Schoen, National Correspondent, Federal Republic of Germany IGD 84/A 29

528.32(091)

Grosse, Hansdieter
Die Basis Oberhergheim 1804—1877—1979
Dt. geod. Kommiss., R. B. — Frankfurt/M. (1981) 257. — p. 39—49 : div. fig. 3 tab. 4 lit. div. app. (= Mitt. Nr. 162 des Inst. Angew. Geod.)
The base line Oberhergheim 1804—1877—1979 (GER)
AS 12.01.09
Key words: Base line, base line measurement

As part of the scale controls conducted in the West German part of the European Primary Triangulation Net since 1958, a control survey of the base line Oberhergheim at the connection of the German-French networks was carried out in 1979. A discrepancy of 32 cm between the base line measurements of 1804 and 1876, the results of which have not been introduced repeatedly in the calculations of the first order nets, could be cleared up.

The checking of the length of this base line is interesting mainly for historical reasons.

Schoen, National Correspondent, Federal Republic of Germany IGD 84/A 30

528(091)

Izotov, A. A.
Ob izuchenii istorii geodezii kak otrasli nauki i tekhniki
Geod. i Kartogr. — Moskva (1982) 8. — p. 28—35 : 7 lit.
On the study of the history of geodesy as a branch of science and technology (RUS)
AS 12.01.09
Key words: History of geodesy

Dobrokhotov, National Correspondent, USSR IGD 84/A 31

528.8:377/378
681.32:528.711.7

Jensen, John R.
Educational image processing: an overview (ENG)
Photogramm. Engin. & Remote Sens. — Falls Church, Virginia 49 (1983) 8. — p. 1151—1157 : 1 tab. div. lit.
AS 12.01.79/12.29.05/63.11
Key words: Digital computer, digital image processing, education

Digital image processing is often an integral part of a student's education in remote sensing. Such instruction is normally accomplished by providing digital image processing laboratory assignments within the context of introductory remote sensing courses and/or teaching separate classes or seminars specifically oriented toward digital image processing of remotely sensed data. The general course content of such approaches is briefly reviewed. In addition, image processing system configurations to support these scenarios are identified and discussed.

(Author's summ.)

Techn. Univ. Dresden IGD 84/A 32

528.48

Milev, G.
Normirane, standartizacija i sistemi ot dopuski i dopustimi otklonenija pri geodezicheskite raboti v stroitelstvoto
Geod., Kartogr., Zemeustrojjstvo. — Sofija (1983) 2. — p. 11—14 : 5 fig. 1 tab. 8 lit.
Normalization, standardization and systems of admissible tolerances for geodetic operations in civil engineering (BUL)
AS 12.01.37/12.23.05/12.23.09
Key words: Engineering geodesy, standardization

Georgiev, National Correspondent, Bulgarian IGD 84/A 33

528.4-057(492)(091)

Muller, E.
Nederlandse landmeters en hun opleiding voor 1800
Spiegel Historiael. — Haarlem 16 (1981) 9. — p. 450—455 and 510 : 8 fig. 5 lit.
Dutch landsurveyors and their schooling before 1800 (DUT)
AS 12.01.79
Key words: Landsurveyor, the Netherlands

Starting in the middle-ages the author gives a description of the schooling of land-surveyors in the Netherlands. Before 1500 this schooling was mainly a practical one.

During the 16th century, surveying became a science due to the work of Gemma Frisius and the progress of mathematics e.g. the introduction of the decimal system by Simon Stevin.

At the universities of Leiden and Franeker and other schools lessons in surveying were given.

Attention is paid to the various schools and the importance of the art of fortification.

Pouls, National Correspondent of the Netherlands IGD 84/A 34

528(430.1)
92 Nittinger

Nittinger, Johannes
Einige Lehren aus einem 50jaehrigen geodaetischen Berufsleben
Allg. Vermess.-Nachr. — Karlsruhe 90 (1983) 8—9. — p. 358—366 : 4 lit.
Some lessons from a 50-year geodetic professional life (GER)
AS 12.01.09
Key words: Nittinger, survey in the Federal Republic of Germany

Techn. Univ. Dresden IGD 84/A 35

528.44(492)(091)

Scheffer, A.
Het Kadaster in de provincie Noord-Braband van 1811 tot 1844
Geodesia Nederl. geod. T. — Apeldoorn 24 (1982) 4. — p. 117—126 : 7 fig. 4 lit.
The Cadastral Service in the Netherlands province of North-Brabant 1811—1844 (DUT)
Part 1
AS 12.01.09
Key words: Cadastre, the Netherlands

A historical description of the beginning of the cadastre and the work done by cadastral surveyors.

(Author's summ.)

Pouls, National Correspondent of the Netherlands IGD 84/A 36

06.05:528(430.2)

Sieber, Gerd
Zum Beitrag der Wissenschaftlich-Technischen Gesellschaft fuer Geodaesie, Photogrammetrie und Kartographie fuer die intensiv erweiterte Reproduktion im staatlichen Vermessungs- und Kartenwesen
Vermessungstechnik. — Berlin 32 (1984) 1. — p. 2—4
The contribution of the Scientific-Technological Society for Geodesy, Photogrammetry and Cartography to the intensively extended reproduction in state surveying and mapping organization (GER)
AS 12.01.13
Key words: GDR, Scientific-Technological Society for Geodesy, Photogrammetry and Cartography

According to the directions of the presidency of the Chamber of Technology (KDT) for the preparation of the VIII Congress of the KDT the obtained results of socialist collective work in the framework of the Scientific-Technological Society for Geodesy, Photogrammetry and Cartography (WTG GKP) are estimated from the point of view of state surveying and mapping organizations and suggestions for further activities of the WTG GKP are given.

(Author's summ.)

Techn. Univ. Dresden IGD 84/A 37

528.3(430.1)(091)

Wolf, Helmut
Die deutsche Landesvermessung in ihrer historischen Entwicklung und ihren kuenftigen Moeglichkeiten
Beitr. Festschr. 100 Jahre geodaetische Lehre und Forschung in Hannover Wiss. Arb. Fachricht. Vermess.-Wesen Univ. Hannover. — Hannover (1981) 100. — p. 47—53
German national survey: its historical development and future possibilities
Contribution from the commemorative publication "100 years of geodetic science and research in Hanover" (GER)
AS 12.01.09/12.15.25
Key words: Germany, national survey

Division:

1. From C. F. Gauss to Schreiber
2. The Schreiber era
3. The German standard system of 1923
4. The triangulation network of the German "Reich"
5. The German primary triangulation network
6. Net adjustment on a European scale
7. Future possibilities of development in German national survey.

Schoen, National Correspondent, Federal Republic of Germany IGD 84/A 38

Technical University of Dresden
Section Geodesy and Cartography

bibliographia geodaetica

AIDOS Systematics **12.15** **GROUP I GEODESY**

Geodesy, geodetic measuring techniques and their treatment

GRUPPE I HÖHERE GEODÄSIE

Höhere Geodäsie, geodätische Meßverfahren und ihre Bearbeitung

AIDOS Systematics **12.15.05** **Group I Geodesy**

Mathematical geodesy

Co-ordinate determination, co-ordinate transformation, reference bodies and systems used for computing processes
The Earth as a sphere, a spheroid, an ellipsoid . . .
Mapping of the Earth body on reference surfaces
Relations between the figure of the Earth and reference surfaces
Solving of geodetic main problems

Gruppe I Höhere Geodäsie

Mathematische Geodäsie

Koordinatenbestimmung, -transformation
Für Rechenprozesse verwendete Körper und Systeme
Erde als Kugel, Sphäroid, Ellipsoid . . .
Abbildung des Erdkörpers auf Näherungsflächen
Beziehungen zwischen Erdfigur und Näherungsflächen
Lösung geodätischer Hauptaufgaben

528.22:528.247

Bjerhammar, Arne
A stochastic approach to the mixed boundary value problem in physical geodesy (ENG)
Royal Inst. Technol., Dept Geod. — Stockholm 1983. — 18 p.
AS 12.15.05/12.15.09
Key words: Bjerhammar's method, boundary value problem, mathematical geodesy, Minque, physical problem

The mixed boundary value problem with gravity anomalies over continents and radar altimeter data over oceans is given in a stochastic solution by harmonic reduction down to an internal sphere. The proper weighting between the two observables is achieved by the Minque technique.

Sjoeberg, National Correspondent, Sweden IGD 84/I 151

528.23:528.335

Crosilla, Fabio
A criterion matrix for the Second Order Design of control networks (ENG)
Bull. geod. — Paris 57 (1983) 3. — p. 226–239 : 4 fig. 16 lit.
AS 12.15.05/12.15.25
Key words: Criterion matrix, design of geodetic network

This paper proposes an original method of constructing a criterion matrix for the optimal design of control networks be means of the contraction of the eigenvalues and the rotation of the eigenvectors of a covariance matrix. The rotation of the eigenvectors is carried out by two different methods, either by independent rotations of the pairs of the eigenvector matrix components or by a procrustean transformation of the same matrix. The required criterion is that the "essential eigenvector components" must be as orthogonal as possible to the predicted direction of deformation. A Second Order Design problem is then resolved, which is the optimization of the precision of the observations of a local free distance network to be constructed for the study of recent crustal movements in the seismogenetic area of Friuli (Italy). The results obtained show a high degree of correspondence for the two proposed methods of rotation if limited rotations of the eigenvectors are considered.

(Author's summ.)

Techn. Univ. Dresden IGD 84/I 152

528.232.1:621.396.96

Czarnecki, Kazimierz
The position and orientation of a reference ellipsoid by Doppler satellite measurements (ENG)
Nabljudenija iskusstv. sputnikov Zemli. — Moskva (1978/1979) 18. — p. 435—442
Rzh 1982/5.52.74
AS 12.15.05/12.15.09/12.15.37
Key words: Computations on the ellipsoid, Doppler observation, geometrical satellite geodesy, transformation of co-ordinates

A network of few control points is to be established, the co-ordinates of which will be determined by translocation. At an accuracy of relative position determination of 1 m the method proposed is as effective and accurate as classical triangulation.

Kombinat Geod./Kart. IWT, Leipzig IGD 84/I 153

528.235:528.9

Dermanis, Athanasios / Livieratos, E. / Paraschakis, I.
Applications of strain criteria in cartography (ENG)
Bull. geod. — Paris 57 (1983) 3. — p. 215—225 : 3 fig. 1 tab. 16 lit.
AS 12.15.05/12.33.13
Key words: Map projection, strain parameter

Some of the most commonly used map projections are analysed from the viewpoint of the deformation introduced when the objective spherical surface is mapped into a plane. After a short theoretical account, formulae for the computation of dilatation and maximum shear strain are given for various projections. These two invariant quantities may be used as criteria for the choice of a map projection in addition to the traditional criteria of conformality, equivalence, etc.

(Author's summ.)

Techn. Univ. Dresden IGD 84/I 154

528.232.1/.2

Ehlert, Dieter
Beziehungen zwischen den Ellipsoidparametern
Nachr. Kt.- u. Vermess.-Wesen, R. I. — Frankfurt/M. (1983) 91. — p. 11—23 : 3 lit.
Relations between ellipsoid parameters (GER)
AS 12.15.05
Key words: Ellipsoidal parameter, mathematical geodesy

The mathematical relationships between the commonly used ellipsoidal parameters are represented by closed formulae and Taylor series.

(Author's summ.)

Techn. Univ. Dresden IGD 84/I 155

528.236:528.14

Gruss, Johannes
Ebene Helmert-Transformation mit gemessenen Netzelementen
Z. Vermess.-Wesen. — Stuttgart 108 (1983) 10. — p. 446—453 : 1 fig. 2 lit.
Planar Helmert transformation with measured net elements (GER)
AS 12.15.05/12.15.37
Key words: Helmert-transformation

Description of formulae for a planar transformation system.

Techn. Univ. Dresden IGD 84/I 156

061.3(450.253.11):528.23

Heck, B.
VIII. Hotine-Symposium ueber Mathematische Geodaesie, Como, Italien, 7.—9. September 1981
Allg. Vermess.-Nachr. — Karlsruhe 88 (1981). — p. 444—446
VIIIth Hotine symposium on mathematical geodesy, Como, Italy, 7—9 September 1981 (GER)
AS 12.15.05/12.15.09
Key words: Hotine symposium, mathematical geodesy, physical geodesy

Under the sponsorship of the IAG the 8th symposium on mathematical geodesy took place at the Villa Olmo in Como, Italy.

At the 1st session, recent developments in geodetic research were discussed. Certain problems connected with the determination of the field of gravity were the main topic of the 2nd session. The 3rd session dealt primarily with numerical problems in connection with approximation and interpolation problems as well as the processing of very large data quantities. The topics of the last session were "falsely formulated" problems and boundary value problems in physical geodesy.

(Author's summ.)

Schoen, National Correspondent, Federal Republic of Germany IGD 84/I 157

528.23
528.34:516.3

Kirillov, V. G.
O konformnosti trekhmernykh otobrazhenijj
Geod., Kartogr. i Aehrofotos"jomka. — L'vov (1982) 36. — p. 56—59 : 7 lit.
On conformity of three-dimensional representations (RUS)
AS 12.15.05/12.15.37
Key words: Mathematical geodesy, three-dimensional space

The proof of orthogonality of Jacobian matrix of conformal transformation in the three-dimensional space consisting of partial derivatives of the components of a new co-ordinate basis with respect to the components of the initial one is given. As stated, such a transformation may be described by polynomials of 2nd and higher degrees.

Dobrokhotov, National Correspondent, USSR IGD 84/I 158

528.27:528.2(26):621.396.969.11
517.946.9:528.2

Sacerdote, F. / Sanso, Fernando
A contribution to the analysis of the altimetry-gravimetry problem (ENG)
Bull. geod. — Paris 57 (1983) 3. — p. 257—272 : 2 tab. 7 lit.
AS 12.15.05/12.15.17/12.15.37
Key words: Altimetry, boundary value problem, gravimetry

The altimetry-gravimetry problem in its modified form, as defined by Pellinen (1980) and Holota (1982) (Δg given on land and δg given on sea), is analysed in spherical approximation: broader results concerning existence and uniqueness than those previously known are achieved. A proposal for the numerical solution of the problem is presented.

(Author's summ.)

Techn. Univ. Dresden IGD 84/I 159

528.235.1
528.94:656.61.052(211)

Zickwolff, G.
The advantages of stereographic projection: charts 2812 and 2813 of the "Deutsches Hydrographisches Institut (DHI)" (ENG)
Internat. hydrogr. Rev. — Monaco 59 (1982) 1. — p. 119—130 : 6 fig. 2 tab.
AS 12.15.05/12.15.21/12.33.41
Key words: Marine geodesy, mathematical geodesy, stereographic projection, thematic chart

Charts 2812 and 2813 show the Northern and Southern hemisphere of the Earth respectively in polar stereographic projection, the chart face being restricted to the approximate coastlines. The charts may be used for global representations of all kinds, for example for hydrographic and meteorological purposes. The peculiarity of stereographic projection is, however, shown to particular advantage if it is intended to represent circles, since circles on the Earth's surface also appear on the chart as circles (or straight lines) if it is permissible to regard Earth approximately as a sphere.

Schoen, National Correspondent, Federal Republic of Germany IGD 84/I 160

AIDOS Systematics **12.15.09** **Group I Geodesy**

Physical geodesy

Theory and methods for exploring the figure and the gravity field of the Earth
Constants of the Earth, physics of the Earth
Movements of the Earth's crust, geotectonics
Geodynamic investigations, gravity measurements
Gravimetry for potential-theoretical foundations

Gruppe I Höhere Geodäsie

Physikalische Geodäsie

Theorie und Methoden zur Erforschung der Figur und des Schwerefeldes der Erde
Konstanten der Erde, Physik der Erde
Erdkrustenbewegungen, Geotektonik
Geodynamische Untersuchungen, Schweremessungen
Gravimetrie für potentialtheoretische Grundlagen

550.312

Andreev, A. A.
Izostazija v zone perekhoda ot Aziatskogo materika k Tikhomu okeanu
Geofiz. polja perekhod. zony Tikhookean. tipa. — Vladivostok (1981). — p. 3—7 : 12 lit.
Isostasy in the transition zone from the Asiatic continent to the Pacific Ocean (RUS)
AS 12.15.09
Key words: Isostasy, physical geodesy

Dobrokhotov, National Correspondent, USSR IGD 84/I 161

551.244(94):629.783

Angus-Leppan, P. V. / Allman, J. S. / Sloane, B.
Crustal movement from satellite observations in the Australian region (ENG)
Proc. General Meet. IAG, Tokyo, May 7—15, 1982. — Tokyo (1982). — p. 272
AS 12.15.09/12.15.17
Key words: Crustal movement, physical geodesy, satellite geodesy

The paper outlines the crustal structure in the Australian and Papua New Guinea regions, and explains the need for data on crustal motion and deformation. This paper describes a third project, designed to determine the motion of plates in the Papua New Guinea region. Here the structure and motions are complex, with several minor plates between the major Indo Australian and Pacific Plates. Fifteen observing stations were selected so that there were at least two on each plate, and in April—May 1981 observations were taken to fix their positions using five Doppler satellite receivers in a multi-station configuration. The observations are currently being processed and analysed. It is hoped to repeat the measurements at regular intervals.

Nagasawa, National Correspondent, Japan IGD 84/I 162

528.47

Antonov, Ju. V.
Rezul'taty issledovanijj po razdeleniju slozhnykh anomalijj sily tjazhesti
Geofiz. metody poiskov i razvedki rud i nerud. mestorozhd. — Sverdlovsk (1980) 6. — p. 95—102
The results of studies on separation of complex gravity anomalies (RUS)
AS 12.15.09
Key words: Gravity anomalies

The problems of developing numerical schemes for separation of complex gravity anomalies are discussed. Compact numerical schemes of separation that permit separating the complex anomalies into the local ones just from the observed field without any intermediate steps are obtained. The testing on the models and practical examples gave satisfactory results.

Dobrokhotov, National Correspondent, USSR IGD 84/I 163

528.21:525.11
517.52

Arnold, Kurt
Die Abschaetzung des Restgliedes der von K. Arnold modifizierten Reihenentwicklung von H. Moritz
Vermessungstechnik. — Berlin 32 (1984) 1. — p. 13—16 : 3 fig. 10 lit.
The estimation of the residual term of the series expansion of H. Moritz modified by K. Arnold (GER)
AS 12.15.09/12.15.05
Key words: Bouguer anomaly, residual term

The series expansion of H. Moritz to compute height anomalies is modified by superposition with the potential of visible rock masses. The different terms of this series expansion are small. Presentation of a technique to estimate the amount of the residual term of the modified series expansion. In regions with very changeable Bouguer anomalies only, the amount of the residual term might become critical.

(Author's summ.)

Techn. Univ. Dresden IGD 84/I 164

528.27.081.4-187.4

Beljaev, A. S. / Rozenberg, V. N.
Vlijanie metodiki izmerenijj na rezul'taty vysokotochnojj gravimetricheskojj s"jomki
Razved. geofiz. — Moskva (1981) 93. — p. 129—133
The effect of the method of measurements on high-precision gravity survey data (RUS)
AS 12.15.09/12.15.33
Key words: Gravity measurement
Dobrokhotov, National Correspondent, USSR IGD 84/I 165

528.026:519.218

Bereza, A. D. / Kostrov, A. V. / Rivkin, S. S.
Issledovanie metodov fil'tracii pri gravimetricheskikh izmerenijakh
Fiz.-tekhn. gravimetrija: apparatura i metody izmerenijj. — Moskva (1982). — p. 61—88 : 12 lit.
Examination of filtering methods in gravity measurements (RUS)
AS 12.15.09/12.15.37
Key words: Filtering, gravity measurement
Dobrokhotov, National Correspondent, USSR IGD 84/I 166

621.396.969.11:629.783(261.26)

Brennecke, Juergen
Aufbereitung der Altimeterdaten fuer das SEASAT-1-Nordsee-Altimeterexperiment
Veroeff. Bayer. Kommiss. internat. Erdmess., astron.-geod. Arb. — Muenchen (1980) 40. — p. 104—110 : 4 fig. 2 tab. 10 lit.
Preparation of the altimeter data for the SEASAT-1-North Sea altimeter experiment (GER)
AS 12.15.09/12.15.17/12.15.21
Key words: Altimeter measurement, marine geodesy, physical geodesy, satellite

On 26th June 1978, NASA started the oceanographic research satellite SEASAT-1. SEASAT-1 was conceived as a prototype of a satellite for the permanent observation of the sea surface.
On 10th October 1978 a defect in the power supply system put all instruments out of function. During the short measurement period, however, the sensors furnished a sufficient amount of measurement data.
The author gives a summary of the extent and the arrangement of the data records on the JPL-tapes. Then he describes in short the extensive programme package for the preparation of the data. For the data-editing a programme with additional criteria for the rejection of gross errors in the altimeter measurement has been established — which are outlined in detail.
Schoen, National Correspondent, Federal Republic of Germany IGD 84/I 167

621.396.969.11:629.783(261.26)
528.21.088.3

Brennecke, Juergen
Berechnung von Geoidprofilen aus Altimeterdaten wiederkehrender Subsatellitenpassagen im Rahmen des Seasat-1-Nordsee-Altimeterexperiments
Veroeff. Bayer. Kommiss. internat. Erdmess., astron.-geod. Arb. — Muenchen (1980) 40. — p. 113—126 : 7 fig. 4 tab. 10 lit.
Computation of geoid profiles from altimeter data or recurring sub-satellite passages within the scope of the Seasat-1-North Sea-Altimeter Experiment (GER)
AS 12.15.09/12.15.17/12.15.21
Key words: Altimeter measurement, geoid, marine geodesy, physical geodesy, satellite

The author reports on the first results obtained from the computation of geoid profiles in the North Sea area, based on altimeter data of sub-satellite passages repeated at intervals of 3 days. The geoid undulations were developed into a kernel function series. Before the determination of the serial coefficients, the seasurface topography has been included as correction into the observations by means of adjustment. The influence of the selection of different solution parameters and data on the resulting geoid profile is described. Finally the accuracy of the adjusted geoid profiles is estimated.

Schoen, National Correspondent, Federal Republic of Germany IGD 84/I 168

528.247

Buzuk, V. V. / Kanushin, V. F.
Metody prognozirovanija anomalijj sily tjazhesti
ONTI CNIIGAiK DEP. — Nr. 89gd-D82. — 29 p. : div. fig. 58 lit.
The methods of gravity anomaly prediction (RUS)
AS 12.15.09/12.15.33
Key words: Gravity anomaly, gravity measurement

Dobrokhotov, National Correspondent, USSR IGD 84/I 169

528.21(436)

. . .
Das Geoid in Oesterreich
Graz: 1983. — Geodaet. Arb. Oesterreichs f. d. internat. Erdmessung. — Neue Folge. — Bd. III
The Geoid in Austria (GER)
AS 12.15.09
Key words: Astro-geodetic geoid determination, computation (astro-geodetic levelling, collocation), observation methods (zenith camera, astrolabe), physical geodesy

A collaboration between the Technical Universities of Graz and Vienna and the Federal Bureau of Standards and Surveying enables Austria to present a first version of an astro-geodetic geoid, which covers 4/5 of the country. The accuracy corresponds to $\pm$ 10 cm in height. The measurements and investigations are the contents of this booklet.

(Author's summ., abbr.)

Erker, National Correspondent, Austria IGD 84/I 170

528.21

Hofmann-Wellenhof, Bernhard
Representation of the gravitational potential by multipoles (ENG)
Mitt. geod. Inst. TU Graz. — Graz (1983) 47. — 197 p.
AS 12.15.09
Key words: Direct and inverse problem, potential theory, tripole, quadrupole and multipole potential, regularisation

Chapter 1 following Moritz gives an overview of the determination of the Earth's potential. In chapter 2 the method of regularization is deduced and applied for the tripole and quadrupole potential of a circle and of a closed curve in the plane. In chapter 3 numerical experiments are described. In chapter 4 it is shown how the density function can be discretized with the introduction of finite elements. Chapter 5 contains some instructions for the computer implementation of the direct and of the inverse problem.

(Author's summ.)

Erker, National Correspondent, Austria IGD 84/I 171

528.24:528.088

Xu Houze / Yang Huaibing
Some remarks on the estimation of truncation error in calculation of height anomalies (CHI)
Acta Geodet. et Cartogr. Sinica. — Peking 12 (1983) 2. — p. 95—102 : 10 lit.
AS 12.15.09
Key words: Deflection of the vertical, height anomaly, truncation error

In this paper we have estimated truncation error for different cap radius in calculation of height anomalies and vertical deflections based on the anomaly degree variance model of Tscherning-Rapp. It is found that truncation error will be rapidly decreased by means of method of modifying the integral kernel. In order to decrease errors two simple and convenient methods for modifying the integral kernel are proposed.

(Author's summ.)

Techn. Univ. Dresden IGD 84/I 172

061.3(100)"1983"(430-2.2)IAG:551.241

. . .
International Association of Geodesy Commission on Recent Crustal Movements (CRCM)
International Center on Recent Crustal Movements (ICRCM) (ENG)
Res. Inst. Geod., Topogr., Cartogr. Praha. — Praha (1983) Bull. 19. — 84 p. (Ed.)
AS 12.15.09
Key words: Crustal movement, XVIII. General Assembly of IUGG Hamburg 1983, physical geodesy

Contents:

1. Introduction
2. Acknowledgement and address
3. Minutes of the CRCM Executive Committee meeting
4. Minutes of the CRCM Commission meeting
5. Reports of the CRCM subcommissions and working groups
6. Scientific communication

Techn. Univ. Dresden IGD 84/I 173

061.3.053.52(430-2.77)"1980"

. . .
4th International Symposium Geodesy and Physics of the Earth, Karl-Marx-Stadt, GDR, May 12th—17th, 1980. Proceedings (ENG)
(Ed. Zentralinstitut fuer Physik der Erde, Potsdam)
Potsdam, Veroeff. Zentralinst. Phys. Erde. (1981) 63, Part I—III. — 836 p. : div. fig. div. tab. div. lit.
AS 12.15.09
Key words: Physical geodesy, symposium

The scientific programme prepared by members of the Scientific Committee included contributions to the following items

— Global and planetary dynamics of the Earth
— Figure and gravity field of the Earth
— Geodynamical processes

which are kept in the structure of the proceedings.

Schoen, National Correspondent, Federal Republic of Germany IGD 84/I 174

528.21

Ivanov, S. S.
Ehvoljucija figury i gravitacionnogo polja svobodno vrashhajushhejjsja planety v processe plotnostnojj differenciacii
Probl. teor. geodinam. i tekton. litosfer. plit. — Moskva (1981). — p. 106—110; 179—180 : 15 lit.
Evolution of the figure and the gravitational field of a freely rotating planet in the process of density differentiation (RUS)
AS 12.15.09/12.15.13
Key words: Earth's ellipsoid, gravity anomaly, potential of the Earth

Based on a numerical model of density differentiation the evolution of angular rotation velocity, semiaxes values and flattening of the Earth's ellipsoid as well as the normal gravitational potential on Earth's surface and gravity values at the pole and the equator during the last 4.6 milliard years is discussed. The computations are carried out in approximation to the 2nd-order values with respect to the flattening assuming exterior forces being absent. The variation of the inertia momentum of the Earth results in attenuation of the dependence of flattening and rotational velocity.

Dobrokhotov, National Correspondent, USSR IGD 84/I 175

528.223(234.53)(261.2):551.3

Kanngieser, Erich
Anlage des Geodynamik-Profils Harz-Nordsee
Allg. Vermess.-Nachr. — Karlsruhe 90 (1983) 8—9. — p. 301—307 : 2 fig. 1 tab. 26 lit.
The establishment of the geodynamic profile Harz Mountains-North Sea (GER)
AS 12.15.09
Key words: Earth's crust, geodynamics

By the establishment of the geodynamic profile Harz Mountains-North Sea the pre-conditions were created to study recent vertical crustal movements in Northern Germany by means of modern geodetic measuring methods. Apart from precision levelling and precision gravity measurements with LaCoste-Romberg gravimeters, considerable progress for the determination of variations in time are expected especially from the planned application of the absolute gravimeter of Faller, U.S.A. Furthermore, the established profile enables the calibration of LaCoste-Romberg gravimeters as well as the investigation of hydrological effects on gravity measurements.

Techn. Univ. Dresden IGD 84/I 176

551.241(491.1):519.216.3

Kanngieser, Erich
Modellierung vertikaler Krustenbewegungen durch Kollokation
Z. Vermess.-Wesen. — Stuttgart 108 (1983) 9. — p. 373—381 : 7 fig. 23 lit.
Modelling of vertical crustal movements by collocation (GER)
AS 12.15.09/12.15.37
Key words: Collocation, Earth's crust movement, gravity measurement, levelling

The proposed technique utilizes least-squares collocation for producing a common and continuous model of vertical crustal movements. The data used are the results of repeated levellings and precise gravity measurements. A position-time system is the reference for the sample points, where height and/or gravity have been observed. The necessary covariance functions have been determined empirically, analyzed regarding isotropy and used for the prediction of the variations. The developed algorithm has been applied successfully to data obtained in a geodynamic testfield in Iceland.

(Author's summ.)

Techn. Univ. Dresden IGD 84/I 177

528.247

Kanushin, V. F.
Osnovnye principy prognozirovanija anomalijj sily tjazhesti s uchjotom dopolnitel'nojj informacii
ONTI CNIIGAiK, DEP. — Nr. 90 g d-D 82. — 33 p. : div. fig. 37 lit.
Basic principles of gravity prediction using additional information (RUS)
AS 12.15.09
Key words: Gravity anomaly, gravity field prediction

The conditions that the additional information must satisfy when predicting free air gravity anomalies are formulated. The data on the relief of the Earth are shown to satisfy these conditions most completely. Based on recent geological, geophysical and gravity data and theories on the internal structure of the Earth the main stages of free air gravity field formation and the nature of its geometry on the Earth's surface are discussed. It is shown that isostatically non-compensated topographic masses manifest themselves in the high-frequency band of power spectrum of the anomalous gravity field. Basic principles of gravity anomaly prediction are formulated.

Dobrokhotov, National Correspondent, USSR IGD 84/I 178

528.21:511.7

Keller, Wolfgang / Falk, Reinhard
Lokale Approximation des Quasigeoids aus astrogeodaetischen Daten
Vermessungstechnik. — Berlin 32 (1984) 1. — p. 17–19 : 8 fig. 1 tab. 4 lit.
Local approximation of the quasi-geoid from astro-geodetic data (GER)
AS 12.15.09/12.15.13
Key words: Geodetic astronomy, physical geodesy, quasi-geoid

By linearization and following smoothing a local approximation of the Molodensky problem in gravity space is constructed. This problem may be solved in a closed way. The solution permits a series expansion in terms of cylindrical functions. The gradient of this solution approximates locally the quasi-geoid. The suggested procedure is tested with a model example.

(Author's summ.)

Techn. Univ. Dresden IGD 84/I 179

551.24:528.27

Krasovskijj, S. S.
O projavlenii dinamiki zemnojj kory v gravitacionnom pole
Anomalii geomagnit. polja i glubin. stroenie zem. kory. Materialy Mezhdunar. rab. soveshh. Issled. anomalijj geomagnit. polja i vekov. variacijj, Kiev, 29 okt. — 2 nojab. 1979. — Kiev (1981). — p. 107–111 : 16 lit.
On manifestation of the dynamics of the Earth's crust in the gravity field (RUS)
AS 12.15.09
Key words: Dynamics, Earth's crust movement, gravity field

Dobrokhotov, National Correspondent, USSR IGD 84/I 180

528.247

Lachapelle, Gerard / Mainville, A. / Schwarz, Klaus-Peter
Empirical investigations on the anisotropy of the free air gravity anomaly covariance function (ENG)
Energy, Mines and Resources Canada, Surveys and Mapp. Branch. — Ottawa (1981). — p. 115—131 : 1 fig. 4 tab. 10 lit.
AS 12.15.09
Key words: Gravity anomaly, physical geodesy

Empirical determinations of directional, i.e. meridional and longitudinal, gravity anomaly covariance functions in mountainous and non-mountainous parts of North America were separately evaluated in order to study the isotropy assumption of gravity anomaly covariance functions used in the application of collocation to the estimation of disturbing potential components. Over 500,000 measured land gravity anomalies covering 55 sample regions were used for this purpose. The isotropy condition is fulfilled to a high degree of confidence in both mountainous and non-mountainous parts. Referencing gravity anomalies to a set of geopotential coefficients helps achieving a still higher degree of isotropy and homogeneity in non-mountainous areas while the same effect in mountainous areas is achieved by removing the linear height correlation from the anomalies.
(Author's summ.)
Techn. Univ. Dresden IGD 84/I 181

528.271:528.247

Lachapelle, Gerard / McConnell, Ken
On the vertical reference system of the Canadian national data base (ENG)
Energy, Mines and Resources Canada, Surveys and Mapp. Branch. — Ottawa (1981). — p. 133—148 : 5 fig. 1 tab. 4 lit.
AS 12.15.09
Key words: Gravity anomaly, physical geodesy

The effect of the departure of the vertical reference system used to derive free-air anomalies from a single equipotential surface in various geodetic applications is described. The departure of the vertical reference system used to calculate land free-air gravity anomalies of the Canadian national gravity data base from a single equipotential surface, is investigated by comparing heights used to derive gravity anomalies with more recent and presumably better heights at some 1581 gravity anomaly points coinciding with lake surfaces and distributed over large areas of Canada. Analysis of the results reveals statistically significant departures of up to 1 m.
(Author's summ.)
Techn. Univ. Dresden IGD 84/I 182

528.247

Laskin, V. M.
O privedenii anomalijj sily tjazhesti k prjamolinejjnomu profilju
Geofiz. metody poiskov i razvedki mestorozhd. nefti i gaza. — Perm' (1981). — p. 62—65 : 2 lit.
On reduction of gravity anomalies to a rectilinear profile (RUS)
AS 12.15.09/12.15.33
Key words: Gravity anomaly, gravity measurement
Dobrokhotov, National Correspondent, USSR IGD 84/I 183

528.21:528.27

Lordkipanidze, G. P.
Ob uravnenijakh sily tjazhesti
Sakartvelos. SSR Mecnierebata Akad. moambe. Soobsh. AN GruzSSR. — (1981) 2. — p. 345—348
On gravity equations (RUS)
AS 12.15.09
Key words: Gravity variation

The formulas for transformation and reduction of the gravitation of a homogeneous sphere under the conditions:
1. gravity is directed normally to the ellipsoid of revolution at the point
2. gravity is directed along the normal gravity corresponding to the attraction of a homogeneous ellipsoid of revolution,

are derived. The formulas are developed for planar and three-dimensional cases but do not account for considerable gravity variations of local character.
Dobrokhotov, National Correspondent, USSR IGD 84/I 184

528.21:528.247

Marych, M. I.
Ob opredelenii vneshnego gravitacionnogo potenciala Zemli
Geod., Kartogr. i Aehrofotos"jomka. – L'vov (1982) 36. – p. 68–74 : 4 lit.
On determination of the external gravitational potential of the Earth (RUS)
AS 12.15.09/12.15.05
Key words: Gravity anomaly, physical geodesy, potential of the Earth

The first approximation of the Molodensky problem solution by a small parameter based on Stokes integral with the gravity anomalies on the sphere formally given as a Taylor series in the degrees of relief heights is derived. For the general case the method consists in determination of the potential for an outer sphere and its analytical continuation to the remaining part of the outer space of the Earth which is not accounted for by the traditional Stokes formula. As the geopotential is independant of the radius of the Earth's sphere, it may be concluded that the method discussed cannot be brought to the method of analytical continuation of gravity anomalies onto an interior sphere and that the solution given is mathematically rigorous. The validity of the formulas is confirmed by their substitution into the boundary-value condition of the problem.
Dobrokhotov, National Correspondent, USSR IGD 84/I 185

528.21:523.4

Meshherjakov, G. A. et al.
Nekotorye voprosy interpretacii gravitacionnogo polja Zemli v aspekte sravnitel'nojj planetologii
Teorija i metod. interpret. gravimagnit. polejj. Dokl. Vses. seminara, Kiev, 1980. – Kiev (1981). – p. 335–340 : 14 lit.
Some problems of interpretation of the gravitational field of the Earth from the point of view of comparative planetology (RUS)
AS 12.15.09/12.15.13
Key words: Gravity field of the Earth, planetary anomaly

The problems of interpretation of planetary anomalies of Earth, Mars and Moon from the point of view of isostatic compensation and density inhomogenity distribution in the mantle of the planet are discussed. The differences of elastic stress caused by anomalous crustal thickness of the planets are estimated and their interpretation is given. Based on the supposition that the lower mantle does not take part in formation of planetary anomalies a density model of the tectonosphere from degree momenta of density in spherical or ellipsoidal layers may be constructed.
Dobrokhotov, National Correspondent, USSR IGD 84/I 186

528.565.089.6

Nakagawa, Ichiro et al.
Precise calibration of scale values of LaCoste and Romberg gravimeters and international gravimetric connection along the circum Pacific zone (ENG)
Proc. General Meet. IAG, Tokyo, May 7–15, 1982. – Tokyo (1982). – p. 298–308 : 3 fig. 2 tab.
AS 12.15.09/73.19
Key words: Calibration, gravimeter, gravity measurement

(s. a. IGD 83/737–740)
Nagasawa, National Correspondent, Japan IGD 84/I 187

528.27(26)

Procaenko, S. V. / Koneshov, V. N.
K voprosu podgotovki apriornojj informacii dlja vypolnenija morskojj gravimetricheskojj s"jomki
Fiz.-tekhn. gravimetrija: apparatura i metody izmerenijj. — Moskva (1982). — p. 139—148 : 4 lit.
The problem of preparation of *a priori* information for marine gravity surveys (RUS)
AS 12.15.09/12.15.21
Key words: Gravity measurement, marine geodesy

Dobrokhotov, National Correspondent, USSR IGD 84/I 188

528.27(57):525.622

Sarycheva, Ju. K. / Timofeev, V. Ju.
Issledovanie prilivnykh variacijj sily tjazhesti v Sibiri
Metodika i rezul. kompl. geofiz. issled. zem. kory Sibiri. — Novosibirsk (1981). — p. 64—70
Determination of tidal gravity variations in Siberia (RUS)
AS 12.15.09/12.15.21
Key words: Gravity measurement, tides

Dobrokhotov, National Correspondent, USSR IGD 84/I 189

528.563(26)

Segawa, Jiro / Kaminuma, Katsutada / Kasuga, Takashi
A new surface ship gravity meter NIPRORI-1 with a servo accelerometer (ENG)
J. Geod. Soc. Japan. — Tokyo 27 (1981) 2. — p. 102—130
Rzh 1982/5.52.131
AS 12.15.09/12.15.21/73.19
Key words: Gravimeter design

The development started in 1979 and tests of the static gravimeter NIPPORT-1 are reported. The precision of the instrument is 0.1 . . . 0.5 mGal.

Kombinat Geod./Kart. IWT, Leipzig IGD 84/I 190

528.024.8:528.27:528.16

Steeves, Robin R.
Estimation of gravity tilt response to atmospheric phenomena at the Fredericton tiltmetric station using a least squares response method (ENG)
Energy, Mines and Resources Canada, Surveys and Mapp. Branch. — Ottawa (1981). — p. 175—180
AS 12.15.09/12.15.33/12.15.37
Key words: Earth's atmosphere, gravity measurement, tilt observation

Variations in the apparent direction of gravity, with respect to the tilting bedrock, at the Fredericton tiltmetric station have been observed, although somewhat intermittently, since 1974. Gaps in these observations account for over half of their overall time span. These tilt observations contain, among others, variations due to atmospheric pressure and surface temperature induced tilts. This work is concerned with modelling the response of the observed tilt, both for purposes of noise reduction and for understanding the phenomena themselves, to the atmospheric pressure and surface temperature variations, which are also observed at this station.

(Author's summ., abbr.)

Techn. Univ. Dresden IGD 84/I 191

528.27:550.312

Strang van Hees, G. L.
De aardse zwaartekracht
Zenit. — 9 (1982) 12. — p. 523—526 : 5 fig. 8 lit.
The gravity of the Earth (DUT)
AS 12.15.09
Key words: Gravity, physical geodesy

A popular description of gravity, gravity measurements and its meaning for geophysical research.
Attention is paid to measurements on land and at sea.
(Author's summ.)
Pouls, National Correspondent of the Netherlands IGD 84/I 192

528.026:528.247

Tarakanov, Ju. A.
Interpretacija sovokupnosti komponentov gravitacionnogo polja i metody ego izmerenija v dvizhenii
Fiz.-tekhn. gravimetrija: apparatura i metody izmerenijj. — Moskva (1982). — p. 171—193 : 12 lit.
Interpretation of various components of the gravity field and the methods of their measurements from a moving station (RUS)
AS 12.15.09/12.15.33
Key words: Gravity anomaly, gravity field of the Earth

Interpretation of various components of the gravitational field is carried out using the rays that go out from the geocentre and form a solid angle including the anomalous body. Along these rays at different heights and depths the elements of the field may be measured, e.g. on the sea-bottom, in a mine, on the land and the ocean surface, in an aeroplane and in a satellite. Such approach makes it unnecessary to reduce gravity anomalies to the sea-level. The form of the anomalous body is presented by a group of points the co-ordinates and the masses of which are calculated from mass momenta determined from a unique solution. The methods of compensation of support accelerations when measuring the first- and the second-order derivatives of the potential by the systems with measurable geometry and mass are discussed.
Dobrokhotov, National Correspondent, USSR IGD 84/I 193

528.21/.24

Torge, Wolfgang
Resultate und Probleme der Geoidbestimmung
Beitr. Festschr. 100 Jahre geodaetische Lehre und Forschung in Hannover Wiss. Arb. Fachricht. Vermess.-Wesen Univ. Hannover. — Hannover (1981) 100. — p. 71—84 : 5 fig. 2 tab. 38 lit.
Results and problems of geoid determination (GER)
Contribution from the commemorative publication "100 years of geodetic science and research in Hanover"
AS 12.15.09
Key words: Geoid determination
Schoen, National Correspondent, Federal Republic of Germany IGD 84/I 194

528.22/.23:517.56

Tscherning, Carl C.
The role of high degree spherical harmonic expansions in solving geodetic problems (ENG)
Prepared for Sympos. "Improved gravity field estimation on a global basis" General Assembly Internat. Assoc. Geod. (IAG). — Hamburg, August 1983. — p. 1—11 : 2 tab. 22 lit.
AS 12.15.09/12.15.05
Key words: Gravity anomaly, mathematical geodesy, physical geodesy

Spherical harmonic expansions to degree and order 180 for the gravity potential, for the topography and for the rock-equivalent topography have recently become available. Such expansions are extremely valuable when studying various geophysical correlations at wavelengths down to about 100 km.

Despite the large number of coefficients used, the evaluation of derived quantities of geodetic interest may easily be computed by recently developed very efficient computer programs. This makes it possible to use (point) gravity field quantities or heights from which the contribution from the high degree expansions have been subtracted. These "residual" quantities will have empirical covariance functions for which the first zero value is located approximately at spherical distance $\varphi_0 = 7°$ for height anomalies, $\varphi_0 = 45'$ for gravity anomalies and $\varphi_0 = 20'$ for the longitudinal components of the deflection of the vertical.

(Author's summ., abbr.)

Techn. Univ. Dresden IGD 84/I 195

551.241:528.02

Wassef, A. M.
Inferring recent crustal movements from repeated geodetic measurements (ENG)
Res. Inst. Geod., Topogr., Cartogr. Praha. — Praha (1983) Bull. 19. — p. 72—84 : 23 lit.
Invited paper read to the Sympos. on Recent Crustal Movements and Phenomena Assoc. with Earthq. and Volcanism. Gen. Meet. IAG, Tokyo, Japan, May 7—15, 1982
AS 12.15.09
Key words: Crustal movement, Earth's crust, geodetic measurement, geophysics

The validity of deducing the recent movements of the Earth's crust by comparing repeated geodetic measurements depends essentially on the confidence which may be put in the estimates of the accuracy of these measurements.
The shortcomings of some systems of error analysis are indicated, the possibilities of improvement are suggested, and the need for concomitant geophysical data is emphasized.

(Author's summ.)

Techn. Univ. Dresden IGD 84/I 196

528.02-187.4:551.244(94)

Wellman, P.
Recent crustal movement determined from repeated surveying — Results from Southeastern and Southwestern Australia (ENG)
Proc. General Meet. IAG, Tokyo, May 7—15, 1982. — Tokyo (1982). — p. 271
AS 12.15.09
Key words: Crustal movement, physical geodesy

First-order geodetic measurements have been repeated in seismically active areas in southeastern and southwestern Australia. In both areas significant crustal movement is detected. The crustal movement given principal axes of compression parallel to those of in situ stress measurements, and roughly consistent with those of earthquake focal mechanisms. The seismic moment rates calculated from the measured strain rates are higher than those calculated from the historic earthquake records. In southeastern Australia the strain rate is about $50 \times 10^{-9}\,yr^{-1}$ over the whole area investigated, but with a gradually changing principal axis of compression. The rate of change of direction is similar to that found in Japan.

Nagasawa, National Correspondent, Japan IGD 84/I 197

528.35:539.3:517.962
517.962:528.35

Welsch, Walter
Finite element analysis of strain patterns from geodetic observations across a plate margin (ENG)
Proc. General Meet. IAG, Tokyo, May 7—15, 1982. — Tokyo (1982). — p. 270
AS 12.15.09/12.15.25/12.15.37
Key words: Finite element, strain, trilateration measurement

Repeated trilateration measurements of local or regional networks are frequently the basis of geodetic strain analyses. As a rule, the strain field is considered homogeneous for the whole area under investigation or at least for parts of it. However, the treatment of strain as homogeneous over a large area means averaging strains and is often not desired.
To get an insight into the local variety of strain patterns, the area under investigation can be dissected into finite elements, i.e. triangles.
Nagasawa, National Correspondent, Japan IGD 84/I 198

528.026.5:531.35
528.568

Zenkov, V. S. et al.
K voprosu ob izmerenii vtorykh proizvodnykh gravitacionnogo potencial v dvizhenii s pomoshh'ju krutil'nogo majatnika
Fiz.-tekhn. gravimetrija: apparatura i metody izmerenijj. — Moskva (1982). — p. 25—42 : 10 lit.
On measurement of the 2nd-order derivatives of the gravitational potential on a moving platform by a torsion pendulum (RUS)
AS 12.15.09/73.19
Key words: Dynamics, gravity potential, physical geodesy, torsion pendulum

Dobrokhotov, National Correspondent, USSR IGD 84/I 199

528.21(084.3):551.4

Zverev, A. T.
K probleme svjazi onduljacijj geoida s tektonicheskimi strukturami
Izvest. vyss. uchebn. Zaved., Geod. i Aehrofotos"jomka. — Moskva (1982) 4. — p. 32—38 : 7 lit.
On relation of geoid undulations to tectonic structures (RUS)
AS 12.15.09
Key words: Geoid, tectonics, undulation

The results of morphometric and correlational analyses of the maps of real and residual geoids and the relief of the Earth are presented. The maps of correlation of geoidal heights and the relief compiled by using different correlation coefficients are given. The inverse correlation of geoidal heights with tectonic structures has been detected. A multi-layered development of geodynamical processes and related tectonic structures and density inhomogenities is shown.
Dobrokhotov, National Correspondent, USSR IGD 84/I 200

AIDOS Systematics **12.15.13** **Group I Geodesy**

Geodetic astronomy/astrometry

Fundamentals of geodetic astronomy
Astronomic determinations of latitude, longitude and azimuth in geodetic networks
Time determination, time service
Astrometry
— Materials, methods, devices for geodetic-astronomic measurement
— Star catalogues, star co-ordinates
— Earth's rotation, polar motion

Observatories, instruments and methods of astronomic observations
— Prerequisites to carry out geodetic-astronomic measurements
— Instruments and devices for observation
— Radio astronomy

Gruppe I Höhere Geodäsie

Geodätische Astronomie/Astrometrie

Grundlagen der geodätischen Astronomie
Astronomische Breiten-, Längen- unnd Azimutbestimmungen in geodätischen Netzen
Zeitbestimmung, Zeitdienst
Astrometrie
— Materialien, Verfahren, Geräte für geodätisch-astronomische Messungen
— Sternkataloge, Sternkoordinaten
— Erdrotation, Polbewegung
Observatorien, Instrumente und Methoden astronomischer Beobachtungen
— Voraussetzungen für die Ausführung geodätisch-astronomischer Messungen
— Beobachtungsinstrumente und -geräte
— Radioastronomie

528.283:529.7

Butkevich, A. V.
Vychislenie azimuta i popravki khronometra v sposobe Kryzhanovskogo pri nabljudenijakh, nesimmetrichnykh otnositel'no meridiana
Geod., Kartogr. i Aehrofotos"jomka. — L'vov (1982) 36. — p. 11—19 : 19 lit.
Calculation of azimuths and chronometer corrections in Kryzanovsky method for observations non-symmetrical relative to the meridian (RUS)
AS 12.15.13
Key words: Azimuth determination, geodetic astronomy, time measurement

Dobrokhotov, National Correspondent, USSR IGD 84/I 201

529.75

Fell, Dieter
Rechnergesteuerter Zeitvergleich fuer das Zeitsystem der Station Wettzell
Veroeff. Bayer. Kommiss. internat. Erdmess., astron.-geod. Arb. — Muenchen (1981) 41. — p. 31—34 : 4 fig. 1 lit.
Computer-aided time comparison for the time system of the Wettzell station (GER)
AS 12.15.13/63.23
Key words: Geodetic astronomy, time comparison, time system

Schoen, National Correspondent, Federal Republic of Germany IGD 84/I 202

522.1(764)(047.2)
528.517.8.089.6:629.783

Hoppe, Walter
Bericht ueber eine Studienreise zum McDonald Observatorium der Universitaet von Texas vom 10. Juli bis 4. September 1979
Veroeff. Bayer. Kommiss. internat. Erdmess., astron.-geod. Arb. — Muenchen (1980) 40. — p. 65—70 : 3 fig. 5 lit.
Report on a study tour to the McDonald Observatory of the University of Texas from July 10th to September 4th, 1979 (GER)
AS 12.15.13/12.15.17
Key words: McDonald Observatorium

Schoen, National Correspondent, Federal Republic of Germany IGD 84/I 203

523.164:522.21

Kilger, Richard
Untersuchung der spezifischen Steifigkeit eines flach- und eines tiefgewoelbten Parabolspiegels fuer ein Radioteleskop
Veroeff. Bayer. Kommiss. internat. Erdmess., astron.-geod. Arb. — Muenchen (1980) 40. — p. 30—40 : 6 fig.
Investigation of the specific flexural rigidity of a flat and deep-curved parabolic reflector for a radio telescope (GER)
AS 12.15.13/47.51
Key words: Radio-astronomy, radio-telescope

The flat-curved reflector (f/D = 0.45) is specifically more rigid than the deep-curved reflector (f/D = 0.35) and a less serious flexure Δr can be expected from it. The advantage of this computing procedure lies in its simple handling (without computer programme) and in the descriptive form of the computed results — strain, lateral force, moment of flexure and flexure normal to the reflector. The fulfilment of higher expectations such as the recording of the three-dimensional state of tension or the computation of a spatial framework are reserved for the nowadays approved computing procedures, such as the finite-element-method, whose disadvantage consists, however, in the absence of clarity of the computing process.

Schoen, National Correspondent, Federal Republic of Germany IGD 84/I 204

522.21:523.164(234.555)
523.164:681.787

Kilger, Richard
Eine Untersuchung zur Optimierung des fuer Wettzell vorgesehenen Radioteleskops
Veroeff. Bayer. Kommiss. internat. Erdmess., astron.-geod. Arb. — Muenchen (1981) 41. — p. 100—105 : 2 fig. 1 tab. 3 lit.
A study on optimization of the radio telescope planned for Wettzell (GER)
AS 12.15.13/47.51
Key words: Radio-interferometry, radio-telescope, VLBI

In order to evaluate the offers received following the invitation to bid for the radio telescope in September 1980 a catalogue of valuation criteria was drawn up. In this connection the importance of mechanical deformation of the antenna's structure in the error compensation of geodetic-astrometrically oriented very long baseline interferometry (VLBI) was studied.

Schoen, National Correspondent, Federal Republic of Germany IGD 84/I 205

550.312:(523.3/.4+525)

Marchenko, A. N. / Abrikosov, O. A. / Cerklevich, A. L.
Ob opredelenii orientirovki glavnykh osejj inercii planet (Zemlja, Luna, Mars, Venera, Jupiter)
VINITI, DEP. — Nr. 4434-82 Dep. — 48 p. : div. lit.
On determination of orientation of the principal inertia axes of the planets (Earth, Moon, Venus, Mars, Jupiter) (RUS)
AS 12.15.13/12.15.09
Key words: Geodetic astronomy, inertia axes of the planets, physical geodesy

Dobrokhotov, National Correspondent, USSR IGD 84/I 206

061.3(100)"1983"(470):550.3:523
523:550.3

. . .
Mezhdunarodnyjj simpozium «Stroenie i dinamika perekhodnykh zon»
K 100-letiju mezhdunarodnojj planetarnojj geofiziki, Sochi, 7—14 dekabrja 1983 g.
Moskva: Mezhd. Geofiz. Komitet pri Prezid. Akad. Nauk SSSR. 1983. 239 p.
International symposium "Structure and dynamics of transition zones"
In Commemoration of the Centenary of International Planetary Geophysics, Sochi, USSR, December 7—14, 1983 (RUS) and (ENG)
AS 12.15/12.15.13/49.31/12.01.17
Key words: Geophysics, marine geodesy, physical geodesy, planetary geophysics

Abstracts of papers on
— Ocean-continent boundary in time and space: new facts and "old" hypotheses
— Inhomogeneities of the upper mantle relation between the surface structures and the deep structures
— Petrochemical features of transition zones
— Sedimentary basins: regularity in distribution, models of formation and evolution
— Dynamic models of formation and evolution of structures in the active and passive oceanic margins.

Techn. Univ. Dresden IGD 84/I 207

528.28:525.73

Munck, J. C. de
Astronomische refractie in de ionosfeer voor radiogolven van hoge frequentie
Geodesia Nederl. geod. T. — Apeldoorn 24 (1982) 11. — p. 389—392 : 5 fig. 4 tab. 2 lit.
Astronomical refraction in the ionosphere for high frequency radio waves (DUT)
AS 12.15.13/47.51
Key words: Geodetic astronomy, refraction

The astronomical (angular) refraction for radio waves due to the ionosphere is calculated for a simple spherical model of the ionosphere. The refraction increases nearly proportional to the layer thickness and increases with decreasing elevation angle, but it does not become excessive at the horizontal direction.
(Author's summ.)

Pouls, National Correspondent of the Netherlands IGD 84/I 208

528.236:522(98/99)

Drozdzewski, Zbigniew
Obserwacja astronomiczne dla wyznaczen wspolrzednych geograficznych w polskich stacjach polarnych w Arktyce i Antarktyce
Biul. Inform. — Warszawa 27 (1982) 5. — p. 22—26 : 3 lit.
Astronomical observations to determine the geographic co-ordinates at the (Polish) polar stations of the Arctic and the Antarctic (POL)
AS 12.15.13
Key words: Antarctic, Arctic, geographic co-ordinate system, Poland, star catalogue

Description of observations performed in the Institute of Geodesy and Cartography for the elaboration of a star catalogue for different methods for determin-

ing astronomical co-ordinates in polar regions. The tables are published separately.
Kombinat Geod./Kart. IWT, Leipzig IGD 84/I 209

528.28:523.892.1

Pilowski, Karl
Ableitung eines Fundamentalkatalogs aus Zenitkamerabeobachtungen und Bestimmung von absoluten geographischen Oertern geodaetischer Punkte und ihren zeitlichen Veraenderungen
Beitrag aus der Festschrift „100 Jahre geodaetische Lehre und Forschung in Hannover"
Wiss. Arb. Fachricht. Vermess.-Wesen Univ. Hannover. — Hannover (1981) 100. — p. 163—169
Derivation of a fundamental catalogue from observations made with a zenith camera and the determination of absolute geographical positions of geodetic points and their time-related changes
Contribution from the commemorative publication "100 years of geodetic science and research in Hannover" (GER)
AS 12.15.13
Key words: Fundamental catalogue, geodetical astronomy, zenith camera

Reference is made to the publication of the Astronomical Station of the University of Hannover, Nr. 13: "Fundamental catalogue, zenith camera, inertial system, systematics of stellar kinematics", in which the subject is dealt with in depth.
Schoen, National Correspondent, Federal Republic of Germany IGD 84/I 210

521.93

Sabadini, R. / Boschi, E.
A geophysical discussion on the secular instability of the rotation pole (ENG)
Boll. Geof. teor. ed appl. — Trieste 23 (1981) 92. — p. 283—296 : 35 lit.
AS 12.15.13
Key words: Earth mantle, Earth model, pole movement

The periodical expansion and retraction of big ice masses during the last 4 to 5 millions of years may have caused the shift of the rotation axis up to two degrees. This value is obtained by an analysis of palaeomagnetic data.
Kombinat Geod./Kart. IWT, Leipzig IGD 84/I 211

521

Stiefel, E. L. / Scheifele, G.
● **Linear and regular celestial mechanics. Perturbed two-body motion. Numerical methods. Canonical theory** (ENG)
Berlin: Springer-Verl. 1971. 306 p. : div. fig.
AS 12.15.13
Key words: Celestial mechanics

Part I — Basic and numerical theory:
- Regularized theory
- Kepler motion
- The initial value problem
- The fundamental differential equations
- Typical perturbations
- Refined numerical methods

Part II — Numerical theory:
- General and classical canonical theory of the perturbations of elements
- The canonical theory of the oscillator generated by the perturbed problem of two bodies

Part III — Geometry and outlook:
— Geometry of the KS-transformation.
Schoen, National Correspondent, Federal Republic of Germany IGD 84/I 212

528.8

Stiller, Heinz / Moehlmann, Diedrich
Extraterrestrische Astronomie — eine aktuelle Herausforderung
Jenaer Rdsch. — Berlin 27 (1982) 3. — p. 108—109
Extraterrestrial astronomy — a topic challenge (GER)
AS 12.15.13
Key words: Extraterrestrial astronomy

Techn. Univ. Dresden IGD 84/I 213

AIDOS Systematics **12.15.17** **Group I Geodesy**

Space geodesy

Techniques of satellite geodesy for determining the figure of the Earth
Solving geodetic problems with the aid of artificial earth satellites
Geometric and dynamic methods of satellite geodesy
Satellite networks
Satellite observation, methods and means
Observation of the Moon and of distance space objects for geodetic purposes
Laser ranging to the Moon
Very-long-base interferometry

Gruppe I Höhere Geodäsie

Kosmische Geodäsie

Verfahren der Satellitengeodäsie zur Bestimmung der Erdfigur
Lösung geodätischer Aufgaben mit Hilfe künstlicher Erdsatelliten
Geometrische und dynamische Methoden der Satellitengeodäsie, Satellitennetze
Satellitenbeobachtung, Methoden und Mittel
Beobachtung des Mondes und von entfernten kosmischen Objekten für geodätische Zwecke
Laserortung des Mondes
Very-long-base-interferometry

528.331:528.225:629.783
621.396.962

Ashkenazi, Vidal / Preiss, W. J. / Anderson, B.
BRITDOC: A geodetic terrestrial and space network (ENG)
Survey Rev. — Tolworth 27 (1983) 209. — p. 117—124 : 3 fig.
AS 12.15.17/12.15.13
Key words: Doppler measurement, satellite geodesy, VLBI

The original aims of the BRITDOC campaign (British Radio Interferometric Terrestrial and Doppler Campaign) are reported in this article.
Cotter, National Correspondent of Great Britain IGD 84/I 214

528.2:629.783:621.396.96

Boal, J. D. / Vamosi, S.
Comparative evaluation of microprocessor controlled transit satellite Doppler receiving equipment (ENG)
Energy, Mines and Resources Canada, Surveys and Mapp. Branch. — Ottawa (1981). — p. 69—91 : 1 fig. 8 tab. 10 lit.
AS 12.15.17/12.15.09
Key words: Doppler observation, satellite geodesy

In the fall of 1979, initially to determine their sensitivity to radio interference, several makes and models of satellite receiving equipment were tested in the Ottawa area. Through the co-operation of the suppliers of the equipment and other groups, a more ambitious plan was developed to ascertain the accuracy of satellite surveys on short lines in urban areas.
Data reductions were carried out using both the U.S. Navy TRANSIT broadcast and the precise satellite ephemerides supplied by the Defense Mapping Agency Hydrographic Topographic Centre (DMAHTC) in Washington. Abnormally high levels of solar activity during the test period seriously degraded the predicted broadcast orbits and the internal consistency of the observations.
(Author's summ.)
Techn. Univ. Dresden IGD 84/I 215

528.2:629.783:621.396.96

Boal, J. D.
Satellite Doppler surveying (ENG)
Energy, Mines and Resources Canada, Surveys and Mapp. Branch. — Ottawa (1980). — p. 15—48 : 7 fig. 18 lit.
AS 12.15.17/12.15.09
Key words: Doppler observation, satellite geodesy

The use of the Navy Navigation Satellite System NNSS for geodetic surveying is now becoming commonplace. It is important that the surveyor and the user of the observed positions be aware of the various methods which can be employed and the limitations which might be placed on the use of the results.
Although sub-metre relative accuracy can be achieved routinely it is unlikely that the NNSS (or TRANSIT system as it is often called) will ever achieve an accuracy of a few centimetres. The latter accuracies will require the deployment of a new satellite system NAVSTAR/GPS which will exist during the second half of this decade. Meanwhile much useful work can be done with TRANSIT and this paper attempts to outline the factors to be considered in TRANSIT surveying.
(Author's summ.)
Techn. Univ. Dresden IGD 84/I 216

528.2:629.783:681.3

Hajos, T.
Data base for satellite observations (ENG)
Nabljudenija iskusstv. sputnikov Zemli. — Moskva (1978/1979) 18. — p. 605—619
Rzh 1982/5.52.171
AS 12.15.17/63.11
Key words: Data bank, satellite observations, storage organization
Kombinat Geod./Kart. IWT, Leipzig IGD 84/I 217

528.2/.3:629.783:621.396.962

Lachapelle, Gerard / Kouba, Jan
Relationship between terrestrial and satellite Doppler systems (ENG)
Energy, Mines and Resources Canada, Surveys and Mapp. Branch. — Ottawa (1980). — p. 49—64 : 2 tab. 28 lit.
AS 12.15.17/12.15.09
Key words: Doppler observation, satellite geodesy

The NWL9D system, which is used to calculate satellite Doppler positions worldwide, appears to be compatible with CIO pole at the 0.″05 accuracy level. Its longitudes (East) should be increased by 0.″08 (± 0.″1) to make its zero geodetic meridian plane parallel with the zero astronomic meridian plane of the BIH. z co-ordinates should be increased by 4 m (± 1 m) to make its centre coincide with the geocentre.
(Author's summ.)
Techn. Univ. Dresden IGD 84/I 218

528.021.061:629.783

Munck, J. C. de
De ionosfeercorrectie bij afstandmeting naar een buitenaards object
Geodesia Nederl. geod. T. — Apeldoorn 24 (1982) 12. — p. 429—433 : 3 fig. 4 tab. 8 lit.
Ionospheric correction for (pseudo) range measurements to satellites (DUT)
AS 12.15.17/12.15.33/12.15.09
Key words: Cosmic geodesy, physical geodesy, satellite

If range- and doppler measurements to a satellite are corrected for ionospheric range effect by using two radio frequencies, some higher order terms will remain. In this paper these terms are calculated for a simple model of the ionosphere, starting from the refraction theory of Moritz. One of the conclusions is that the Earth magnetic effect is not always negligible.

(Author's summ.)

Pouls, National Correspondent of the Netherlands IGD 84/I 219

522.1:629.783(234.555)

Schneider, Manfred
Konzept der Fundamentalstation Wettzell
Veroeff. Bayer. Kommiss. internat. Erdmess., astron.-geod. Arb. — Muenchen (1981) 41. — p. 16—23 : 3 fig.
Concept of the fundamental station Wettzell (GER)
AS 12.15.17/12.15.09
Key words: Satellite observation station Wettzell

The article begins with an explanation of the concept of a fundamental station Wettzell. Then the actual state of work at Wettzell station is described and it is pointed out what further equipment and constructional measures are necessary to make Wettzell a fundamental station.

(Author's summ.)

Schoen, National Correspondent, Federal Republic of Germany IGD 84/I 220

528.28:629.783
528.341.087.5:629.783

Wolf, Erich
Zum methodischen Beitrag der Richtungsmessungen
Veroeff. Bayer. Kommiss. internat. Erdmess., astron.-geod. Arb. — Muenchen (1980) 40. — p. 46—50 : 2 fig. 5 lit.
On the methodic contribution of direction measurements (GER)
AS 12.15.17/12.15.09
Key words: Direction measurement, planetary geodesy, satellite

The author points out in a concise form some fundamental ideas with regard to this subject. They are based on the present working programme with special emphasis placed on

— determination of the gravitational field of the Earth

— orbit determination of artificial Earth satellites

— determination of station co-ordinates

as well as on the recent trends of development which result principally from the progress already achieved or still to be expected in the measuring technique.

Direction measurement is dealt with in general. However, in the individual chapters the author also points out in particular the special case of photographic direction measurement which is considered a useful intermediary solution.

Schoen, National Correspondent, Federal Republic of Germany IGD 84/I 221

AIDOS Systematics **12.15.21** **Group I Geodesy**

Marine geodesy

Determination of the figure of the Earth below seas and oceans
Positioning on the water surface
Variations of the sea level — oceanic tides
Methods and means of measurements of marine geodesy
Submarine earthquakes and geodetic measurements

Gruppe I Höhere Geodäsie

Meeresgeodäsie

Bestimmung der Figur der Erde unter Meeren und Ozeanen
Lagebestimmung auf der Wasseroberfläche
Schwankungen des Meeresniveaus — Meeresgezeiten
Methoden und Mittel der Messungen der Meeresgeodäsie
Meeresbeben und geodätische Messungen

528.47:629.783
528.024(26):621.396.969.11

Cruz, Jaime
Experiences with altimeter data gridding (ENG)
Columbus: Ohio State Univ., Rep. Dep. Geod. Sci. & Surv. (1983) 347. — 57 p. : 26 fig. 4 tab. div. lit.
AS 12.15.21/12.15.17/12.15.37
Key words: Altimetry, least squares collocation, marine geodesy, prediction, satellite geodesy

Altimeter data gridding is a process of predicting the sea surface heights on a regular grid, based on data values given along the altimeter subsatellite tracks. The gridded form of altimeter data is useful in such applications as the contouring of the sea surface, and the comparison of various data. This report gives actual gridding examples and corresponding contour maps of the sea surface, to illustrate the effect of grid spacing, number and configuration of data points used to predict a grid value, and prediction method used. The prediction methods tested include least squares collocation, Bjerhammar's inverse distance weighting method, Hardy's multiquadric method using a hyperbolic kernel, a general harmonic kernel method using Krarup's kernel, and a piecewise biquadratic surface fitting method.
(Author's summ., abbr.)

Techn. Univ. Dresden IGD 84/I 222

528.475:528.548

Gouic, M. le
L'imagerie du fond avec divers types de sondeurs
Geometre. — Paris 126 (1983) 8—9. — p. 43—51 : 10 lit.
Images of the sea bottom with different types of sounding devices (FRE)
AS 12.15.21
Key words: Marine geodesy, depth sounding

This contribution compares bathymetric and morphological results of two surveys performed in Nov. 1979 with different instruments (classical plummet and multiple-beam sounding device).

Techn. Univ. Dresden IGD 84/I 223

528.338(26)

Hooijberg, M. / Loesberg, P. P.
Plaatsbepaling op zee in de Pulau Tujuh Archipel
Geodesia Nederl. geod. T. — Apeldoorn 24 (1982) 3. — p. 81—85 : 7 fig. 7 lit.
Position-fix at sea in the Pulau Tujuh Archipelago (DUT)
AS 12.15.21/12.23.17
Key words: Marine geodesy, positioning at sea

The paper discusses the application of positioning for legal and technical purposes in the exploration for and mining of non-ferrous mineral resources and the variety of tools used in sub-bottom profiling, drilling and dredging over large sea areas in South-East Asia. At the end of 1979, the world's largest bucket dredge was put into operation in Indonesia with a Dynamic Position System (DPS) on board. Special attention is given to the DPS supplying signals to the winches in such a way that the pre-planned pattern will be dredged automatically as required.

Pouls, National Correspondent of the Netherlands IGD 84/I 224

551.461.2:525.623

Laukart, Wilhelm
Ermittlung langfristiger Wasserstandsaenderungen aus Pegelregistrierungen. Dissertation Univ. Hannover
Wiss. Arb. Fachr. Vermess.-Wes. Univ. Hannover. — Hannover (1981) 104. — 140 p.
Determination of long-term changes in water level using gauge recordings. Doctorate thesis Univ. Hannover (GER)
AS 12.15.21
Key words: Filter theory, marine geodesy, physical geodesy, sea level

Firstly the physical causes for changes in sea level as well as data acquisition and preparation are dealt with.

Following a short summary of the theory of random procedure the filter theory and the problems arising from its practical application are discussed. In particular the questions of time series digitization, data decimation, gap closing etc. are described. By means of a band pass filter the water level curve in the short-period range was analysed spectrally.

The relative rise in water level was computed using the method of least squares.

(Author's summ., abbr.)

Schoen, National Correspondent, Federal Republic of Germany IGD 84/I 225

528.47:629.783
528.024(26):621.396.969.11

Liang, Chih-kuo
The adjustment and combination of GEOS-3 and SEASAT altimeter data (ENG)
Columbus: Ohio State Univ., Rep. Dep. Geod. Sci. & Surv. (1983) 346. — 68 p. : 20 fig. 6 tab. 1 app. div. lit.
AS 12.15.21/12.15.17/12.15.37
Key words: Adjustment, altimetry, marine geodesy, satellite geodesy

The report documents the processing of the revised NOAA-NGS 3.5 year GEOS-3 altimeter data at The Ohio State University. The goal of the processing is to put the GEOS-3 data set in the same system as the adjusted SEASAT data set and make them consistent with each other so that they can be used as a combined data set in related studies.

A new procedure was used to edit GEOS-3 data and to compute ground track crossings (crossovers). The new editing procedure is simpler than both the old GEOS-3 and the SEASAT editing procedure.

The adjusted GEOS-3 sea surface heights were compared with the adjusted

SEASAT data. The combined data were compared with both adjusted GEOS-3 and adjusted SEASAT data.
(Author's summ., abbr.)
Techn. Univ. Dresden IGD 84/I 226

525.623:627.52

Maddux, William S.
Datum extrapolation by simultaneous comparison of partial tidal cycles (ENG)
Survey. and Mapp. — Washington 42 (1982) 2. — p. 139—149 : div. lit.
AS 12.15.21
Key words: Coast protection, height datum, tidal effects

A method for extrapolating local tidal data from short observations through simultaneous comparison of partial tide cycles is described. Mean high water can be determined from observations restricted to times near high water at the control and subordinate stations so that:

1. Local data needed for tidal boundary determinations can be obtained for intertidal areas where lower range observations are not possible.
2. Because full range observations are not required even at the control station, the cost and difficulty of obtaining field data is reduced, particularly where data are to be taken by direct staff observation.

The derivation and technical aspects of the method are discussed and instructions for its application are summarized.
(Author's summ.)
Kombinat Geod./Kart. IWT, Leipzig IGD 84/I 227

AIDOS Systematics **12.15.25** **Group I Geodesy**

Establishing of geodetic control networks

Network design, optimum network configuration
Geodetic-astronomic control points
Measurements and evaluations of geodetic networks

Gruppe I Höhere Geodäsie

Anlage geodätischer Festpunktnetze

Projektierung von Netzen, optimale Netzkonfiguration
Geodätisch-astronomische Fundamentalpunkte
Messungen und Auswertungen geodätischer Netze

528.3(71):528.14

Babbage, G. / McLellan, C. D.
Status of the redefinition and readjustment of national horizontal and vertical geodetic networks and of the national geodetic data base (ENG)
Energy, Mines and Resources Canada, Surveys and Mapp. Branch. — Ottawa (1981). — p. 15—37 : 3 fig. 1 tab. 4 lit.
AS 12.15.25/12.15.37
Key words: Adjustment, Canada, geodetic network

The Canadian primary horizontal network will form part of the NAD83 continental adjustment. A recent preliminary adjustment of this network indicates satisfactory agreement between the terrestrial and satellite components of the network. Integration of secondary horizontal control to the primary is under way, and is mainly the responsibility of the provinces, except in the territories. Work is generally on schedule, with the continental adjustment to be complete in 1984, and integration of secondary control to follow.
The Canadian vertical network is to be readjustment on the proposed 1987 datum. A plan is being developed to prepare the network for adjustment, and field operations are now directed towards that purpose. Vehicles and equipment to motorize

one first-order levelling crew have been ordered. Corrections for refraction and actual gravity will be applied to all levelling.
The positional component of the national geodetic data base is near completion. It is designed such that several agencies can store and update their own data. Specifications for the measured data component of the data base are in the preliminary stages.

(Author's summ.)

Techn. Univ. Dresden IGD 84/I 228

528.3:528.13

Baehr, Heinz-Guenter
Reihenfolgen fuer Messungen und Punkte eines geodaetischen Netzes nach Gesichtspunkten der Genauigkeit und der Zuverlaessigkeit
Allg. Vermess.-Nachr. — Karlsruhe 90 (1983) 10. — p. 384—391 : 11 lit.
Sequences of measurements and points of a geodetic net from the point of view of accuracy and reliability (GER)
AS 12.15.25/12.15.37
Key words: Accuracy, geodetic network, net-design, reliability

Techn. Univ. Dresden IGD 84/I 229

528.3(71):528.14

Beattie, D. S. / Henderson, J. P. / Pinch, M. C.
Evaluation, strengthening and adjustment of geodetic networks in Canada (ENG)
Energy, Mines and Resources Canada, Surveys and Mapp. Branch. — Ottawa (1981). — p. 39—68 : 10 fig. 2 tab. 17 lit.
AS 12.15.25/12.15.37
Key words: Adjustment, Canada, geodetic network

A decade ago a systematic evaluation of the Canadian networks was undertaken in order to identify weaknesses which could be corrected by making additional measurements and to prepare for a continental adjustment in 1984. This paper describes the evaluation process, the types of additional measurements which were made (including Doppler positioning), and the steps which have been taken in preparing the various data types for an overall adjustment by least squares using the Helmert block method.

(Author's summ.)

Techn. Univ. Dresden IGD 84/I 230

528.35.088

Beetz, Heinrich et al.
Kontrollmessungen im Bereich der Station Wettzell
— Stand der Arbeiten 1981 —
Veroeff. Bayer. Kommiss. internat. Erdmess. — Muenchen (1981). — astron.-geod. Arb. 41. — p. 24—30 : 2 fig. 1 tab. 7 lit.
Control measurements in the area of the Wettzell station
— Progress report 1981 — (GER)
AS 12.15.25
Key words: Geodetic network

If changes in the models of the Earth determined by satellite geodesy are to be ascertained and conclusions drawn from coordinate changes of the observation points of the observation station with regard to its surroundings must be ascertained or possible local displacement parameters taken into account by way of corrections in global considerations. Such a control network is being developed at the Wettzell station and is described here.

(Author's summ.)

Schoen, National Correspondent, Federal Republic of Germany IGD 84/I 231

528.335:528.181

Gerasimenko, M. D.
Vzaimosvjaz' parametrov geodezicheskojj seti i sobstvennykh znachenijj matricy normal'nykh uravnenijj
Izvest. vyssh. uchebn. Zaved., Geod. i Aehrofotos"jomka. — Moskva (1982) 5. — p. 38—44 : 7 lit.
Interrelation of the parameters of a geodetic net and the proper values of the normal equation matrix (RUS)
AS 12.15.25/12.15.37
Key words: Geodetic network, normal equation matrix

The method of computation of the variations of proper values and proper vectors of the normal equation matrix depending on variations of the net parameters. i.e. the co-ordinates of the stations to be determined and the weights of measurements, is discussed. It is shown that the formulas derived may be used for optimization of geodetic nets and for study of conditionality of normal equation matrices with variations of the net parameters.

Dobrokhotov, National Correspondent, USSR IGD 84/I 232

528.32./33-187(434.4)

Grosse, Hansdieter / Wolfrum, Otfried
Massstabskontrolle in der Elsass-Lothringischen Dreieckskette 1977—1979
Dt. geod. Kommiss. Veroeff. R. B (1981) 257 (Mitt-Nr. 162 d. Inst. f. Angew. Geod.). — Frankfurt a. M. — 50 p. : div. fig., tab., lit., app.
Scale controls in the chain of triangles of Alsace-Loraine, 1977—1979 (GER)
AS 12.15.25
Key words: Base line measurement, trigonometrical net

Schoen, National Correspondent, Federal Republic of Germany IGD 84/I 233

528.23(4-15+73):629.783

Hajela, D. P.
Accuracy estimates of gravity potential differences between Western Europe and United States through LAGEOS satellite laser ranging network (ENG)
Columbus: Ohio State Univ., Rep. Dep. Geod. Sci. & Surv. (1983) 345. — 67 p. : div. fig. div. tab. div. lit.
AS 12.15.25/12.15.17/12.15.09
Key words: Earth's gravitational field, gravity anomalies, potential theory, vertical datum connection

The presently attainable accuracy of connecting the vertical data of regions, which are separated by oceans, has been investigated. The vertical datum connections are established through geocentric positions of laser tracking stations by estimating the geopotential at these stations, and also utilizing the geopotential differences between stations in each region obtained through levelling. The Western Europe—USA vertical datum connection has been examined in detail. Accuracy estimates of a few other datum connections are also given.

(Author's summ., abbr.)

Techn. Univ. Dresden IGD 84/I 234

528.331(439):551.244

Joo, Istvan
Investigation on horizontal movements of the Carpathian Basin on the basis of the horizontal control (ENG)
Proc. General Meet. IAG, Tokyo, May 7—15, 1982. — Tokyo (1982). — p. 272
AS 12.15.25/12.15.06
Key words: Geodetic network, horizontal movement

During the past 100—120 year period two high-precision horizontal controls were carried out for the area of Hungary. The time while the first horizontal control was brought into completion extended through the second half of the 19th century to the 1930's. The second horizontal control was implemented during the postwar period. The author has made comparative examinations between the two horizontal networks in the course of which he has deduced azimuthal and scale discrepancies from the data obtained from identical points of the two networks. He has also made correlation analysis for the preceding discrepancies and for the duration incidental to those (i.e. time elapsed between the measurements of the two horizontal controls). Supported by the results of the analysis, he draws conclusions and suppositions about the horizontal movements of the Carpathian Basin.

Nagasawa, National Correspondent, Japan IGD 84/I 235

528.066:528.331(430.1)
528.241

Schmidt, Rudolf / Ehlert, Dieter
Punkt-, Lotabweichungs- und Richtungsverzeichnis
(Diagnoseausgleichung des DHDN. Anschlussnetz Wettzell)
Dt. geod. Kommiss., R. B. – Frankfurt/M. (1982) 262. – p. 221–426 : div. tab.
(= Mitt. Nr. 165 Inst. Angew. Geod.)
Index of points, deflections of the vertical and bearings
(diagnostic adjustment of the primary triangulation net of FRG. Connection net in Wettzell) (GER)
AS 12.15.25/12.15.37
Key words: Diagnostic adjustment, triangulation net

Schoen, National Correspondent, Federal Republic of Germany IGD 84/I 236

528.3

Steeves, Robin R.
A note on the optimal design of geodetic networks (ENG)
Energy, Mines and Resources Canada, Surveys and Mapp. Branch. – Ottawa (1981). – p. 167–173 : 4 lit.
AS 12.15.25
Key words: Geodetic network, optimal design

A brief discussion of the second order design problem (i.e. determination of observational accuracy requirements given the geometrical design and co-ordinates accuracy requirements) of a geodetic network is given. A direct (non-iterative) method is described which avoids the use of generalized inverses; it is suggested that more thought be given to this approach which can be based on models for the structure of network covariance matrices.
(Author's summ.)

Techn. Univ. Dresden IGD 84/I 237

528.35:681.3

Tretjak, K. R.
Odin iz metodov proektirovanija setejj trilateracii s pomoshh'ju EhVM
Geod., Kartogr. i Aehrofotos"jomka. – L'vov (1982) 36. – p. 99–103 : 4 lit.
A method of computer-based designing of trilateration networks (RUS)
AS 12.15.25/12.15.37
Key words: Trilateration net

A new method of designing trilateration nets based on approximate co-ordinates of stations and their accuracy is proposed. The method has been realized on the computers of the unified system. It has been tested on the network consisting of 18 stations with the R.M.S. error of station position of 1.5 cm and that of distance measurements 1 cm. The number of super-abundant lines to be measured is 16. The solution turned out to be near to the optimal one.

Dobrokhotov, National Correspondent, USSR IGD 84/I 238

528.33:528.021.6
528.236

Wilkinson, G. A.
Report on the ISS evaluation project 1979 Saskatchewan (ENG)
Energy, Mines and Resources Canada, Surveys and Mapp. Branch. — Ottawa (1980). — p. 99—114 : 2 fig. 7 tab.
AS 12.15.25/12.15.33/12.15.37
Key words: Co-ordinates system, geodetic network, inertial survey system

A network of ISS stations was established in south-central Saskatchewan in 1978 using standard procedures developed by the Geodetic Survey. A second-order EDM traverse was run through part of the ISS network in 1979 with the objective of providing an independent check on the accuracy of the ISS network. Results of the two surveys are compared and analysed. Unexplained systematic differences and the possibility of large refraction effects in the EDM traverse make it impossible to draw conclusions about distortions that may exist in the ISS network. Comparisons of eight computed interstation angles and distances gave maximum differences of 5 arc-sec and 0.59 m respectively, giving some measure of the relative accuracies of co-ordinates of adjacent stations.

(Author's summ.)

Techn. Univ. Dresden IGD 84/I 239

528.3(434.4):528.14

Wolfrum, Otfried
Die Messergebnisse von 1977—1979 und ein Vergleich mit den Messungen von 1877
Dt. geod. Kommiss., R. B. — Frankfurt/M. (1981) 257. — p. 29—37 : 1 fig. 4 lit. div. app. (= Mitt. Nr. 162 d. Inst. Angew. Geod.)
Measuring results of 1977—79 and a comparison with the measurements of 1877 (GER)
AS 12.15.25/12.15.37
Key words: Adjustment, direction measurement

The direction measurements carried out in 1877 by the Royal Prussian National Survey in the Alsace-Loraine chain of triangles are being adjusted together with the distances measured from 1977 to 1979 by the Technical University of Darmstadt. An average scale error in the chain has been shown of + 1.7 mm/km, while the base line Oberhergheim was measured 34 mm or 4.6 mm/km too long.

Schoen, National Correspondent, Federal Republic of Germany IGD 84/I 240

AIDOS Systematics **12.15.33** **Group I Geodesy**

Geodetic measuring methods

Methods and techniques for geodetic measurements of distances, angles, heights
Gravity measurements
Assessment of the precision of geodetic measurements

Gruppe I Höhere Geodäsie

Geodätische Meßverfahren

Verfahren und Technologien für geodätische Messungen von Strecken, Winkeln, Höhen
Schweremessungen
Beurteilung der Genauigkeit geodätischer Messungen

528.021.7.029.672
528.517

Berchtold, E.
Bemerkungen zur Infrarotdistanzmessung
Vermess. Photogramm. Kulturtechnik. — Zuerich 81 (1983) 7. — p. 243—245 : 2 fig.
Remarks on infrared distance measurements (GER)
AS 12.15.33/73.07/69.51
Key words: Calibration, Distomat, infrared distance measurement

The author reports on results of setting-out of galleries by means of distance measurements with the DI 35. He introduces a general calibration function instead of usual addition and multiplication constants and tries a physical interpretation of this function.

Techn. Univ. Dresden IGD 84/I 241

528.517.089.6

Gelhaus, Rolf
Zur Variation der Additionskonstanten elektrooptischer Entfernungsmesser
Beitr. Festschr. 100 Jahre geodaetische Lehre und Forschung in Hannover
Wiss. Arb. Fachricht. Vermess.-Wesen Univ. Hannover. — Hannover (1981) 100. — p. 117—131 : 12 fig. 4 tab. 14 lit.
Variation in additive constants of electrooptical rangefinders
Contribution from the commemorative publication "100 years of geodetic science and research in Hanover" (GER)
AS 12.15.33/73.07
Key words: Additive constant, electrooptical rangefinder

The article deals with investigations on the ranging device of a Zeiss Elta 4. Only the very short ranges up to 580 m were studied, always with one prism only located at the centre of the object to be sighted, so as to create comparable conditions with regard to the effect of phase inhomogeneities.

Schoen, National Correspondent, Federal Republic of Germany IGD 84/I 242

528.026.3

Gershhanok, V. A.
Vychislenie prirashhenijj sily tjazhesti s uchjotom neravenstva intervalov vremeni mezhdu nabljudenijami
Geofiz. metody poiskov i razvedki mestorozhd. nefti i gaza. — Perm' (1981). — p. 50—53 : 6 lit.
Computation of gravity differences taking into account inequal time intervals between observations (RUS)
AS 12.15.33/12.15.09
Key words: Gravity measurement, physical geodesy

Dobrokhotov, National Correspondent, USSR IGD 84/I 243

528.022/.023.088:528.41

Gridnev, S. P.
Issledovanie tochnosti stvornogo sposoba sozdanija geodezicheskogo planovogo s"jomochnogo obosnovanija
Tr. Celinogr. s.-kh. in-t. — (1982) 42. — p. 107—112
An accuracy analysis of the alignment method of establishing horizontal geodetic control for surveys (RUS)
AS 12.15.33
Key words: Alignment

Dobrokhotov, National Correspondent, USSR IGD 84/I 244

528.381(79):551.244

Holdahl, S. R.
Recomputation of vertical crustal motions near Palmdale, California, 1959–1975 (ENG)
Proc. General Meet. IAG, Tokyo, May 7–15, 1982. – Tokyo (1982). – p. 267
AS 12.15.33/12.15.37/12.15.09
Key words: Crustal movement, levelling, physical geodesy

By analysis of levelling data, a large area in Southern California was identified as having uplifted between 1960 and 1974. The uplift zone included 12,000 sqkm with the maximum vertical movement of 25 cm centered near Palmdale. CASTLE (1978) later increased the estimate of uplift at Palmdale to 35 cm. Vertical motion in the Palmdale region has now been reassessed with corrected levelling data.

The new results show 7.5 $\pm$ 4.0 cm of apparent upward movement at Palmdale.

Nagasawa, National Correspondent, Japan IGD 84/I 245

528.026.5:528.565

Izmajjlov, Ju. P. / Markov, G. S.
Issledovanie variacijj ehffekta Ehtvesha
Fiz.-tekhn. gravimetrija: apparatura i metody izmerenijj. – Moskva (1982). – p. 89–109 : 4 lit.
The study of variations of Eoetvoes effect (RUS)
AS 12.15.33/12.15.09/73.19
Key words: Eoetvoes effect, gravimeter, gravity measurement

Dobrokhotov, National Correspondent, USSR IGD 84/I 246

528.02-187:528.1

Kahmen, Heribert
Geodaetische Messtechnik und die Theorie der Zufallsfunktionen
Beitr. Festschr. 100 Jahre geodaetische Lehre und Forschung in Hannover
Wiss. Arb. Fachricht. Vermess.-Wesen Univ. Hannover. – Hannover (1981) 100. – p. 141–148 : 7 fig. 8 lit.
Geodetic measuring technique and the theory of random functions
Contribution from the commemorative publication "100 years of geodetic science and research in Hanover" (GER)
AS 12.15.33
Key words: Geodetic measurement, random function

Using examples, this article describes the importance of the a.m. theory for geodetic measuring technique. In particular, measuring technique in connection with parameter and signal estimation and a priori estimation is discussed.

Schoen, National Correspondent, Federal Republic of Germany IGD 84/I 247

528.021.7:551.510.61

Maslich, D. I. / Khizhak, L. S. / Jaskilka, N. B.
Opredelenie koehfficienta turbulentnosti v tochkakh svetovojj krivojj po rezul'tatam geodezicheskikh i meteorologicheskikh izmerenijj
Geodez., Kartogr. i Aehrofotos"jomka. – L'vov (1982) 36. – p. 78–83
Determination of the turbulence factor at the points of the light curve from geodetic and meteorological data (RUS)
AS 12.15.33
Key words: Geodetic measurement, meteorology, optical refraction

Dobrokhotov, National Correspondent, USSR IGD 84/I 248

528.021.7:551.510.61

Maslich, D. I. et al.
Issledovanie nekotorykh kharakteristik turbulentnogo potoka na puti vizirnogo lucha po meteorologicheskim i geodezicheskim izmerenijam
Geod., Kartogr. i Aehrofotos"jomka. — L'vov (1982) 36. — p. 74—78
Determination of some parameters of a turbulent flow in the path of the sight line from meteorological and geodetic measurements (RUS)
AS 12.15.33
Key words: Geodetic measurement, optical refraction, meteorology

Dobrokhotov, National Correspondent, USSR IGD 84/I 249

528.517.8-182.3:629.783

...
Mobiles Laserentfernungsmess-System
Veroeff. Bayer. Kommiss. internat. Erdmess. — Muenchen (1981) astron.-geod. Arb. 41. — p. 124—126 : 4 fig.
Mobile laser ranging system (GER)
AS 12.15.33/73.03/12.15.17
Key words: Laser distance measurement, satellite

Schoen, National Correspondent, Federal Republic of Germany IGD 84/I 250

528.022.1:528.14

Morozov, S. G. / Abramovich, B. L.
Uravnivanie napravlenijj na iskhodnykh punktakh
Geod. i Kartogr. — Moskva (1982) 9. — p. 15—17
The adjustment of the directions at the initial stations (RUS)
AS 12.15.33/12.15.37
Key words: Adjustment of direction, direction measurement, weight of measurement

Some cases of adjustment of directions at the station for several groups of measurements available are discussed. The formulas and possible deviations of the rigorous adjustment and the simplified one are given. When measuring the directions in several groups it is recommended to connect these groups by at least two common directions. For a computer-based adjustment it is proposed to take into account the weights of measurements.

Dobrokhotov, National Correspondent, USSR IGD 84/I 251

528.024

Penton, C. R.
Methods of vertical mapping control (ENG)
Energy, Mines and Resources Canada, Surveys and Mapp. Branch. — Ottawa (1980). — p. 65—82 : 1 fig. 6 tab.
AS 12.15.33/12.01.75
Key words: Height measurement, vertical mapping control

Spirit levelling offers the highest accuracy, is relatively expensive, but has a low production rate. The inertial survey system gives high accuracy and production rates with low costs. Altimeter traverses employ pressure altitude readings to derive elevations. The method is fast and fairly economical, but has the lowest accuracy. The ground elevation meter has advantages of good accuracy coupled with lowest costs, but has low production and requires passable roadways. Trigonometric surveys, while they can provide a good combination of accuracy, cost and production, require line of sight between observing points and hence have limited application.

(Author's summ., abbr.)

Techn. Univ. Dresden IGD 84/I 252

528.381(73):551.244

Reilinger, R. / Brown, L. / Oliver, J.
Relevelling evidence for crustal deformation in the U.S. (ENG)
Proc. General Meet. IAG, Tokyo, May 7—15, 1982. — Tokyo (1982). — p. 266
AS 12.15.33/12.15.25/12.15.09
Key words: Crustal movement, levelling, physical geodesy

We have reduced and analyzed over 90 % of existing U.S. National Geodetic Survey relevelling observations, in order to evaluate the implications of these measurements for contemporary crustal movements. The data are displayed both in profile form (movement versus distance along route) and, where adequate coverage is available, as maps of apparent elevation change. Our analysis involves:

1) evaluating the reliability of the levelling measurements using modified geodetic checks as well as through comparison with independent measurements of crustal movement (e.g. tide and lake level measurements, seismological evidence, tilt and strain observations)
2) interpreting what appear to be real movements in light of other geological and geophysical information on crustal dynamics.

Nagasawa, National Correspondent, Japan IGD 84/I 253

528.35:528.14

Ribarov, S.
Po vahprosa za tezhestite na izmervanijata pri izravnenie na ahglovo-linejjni mrezhi
Geod., Kartogr., Zemeustrojjstvo. — Sofija (1983) 2. — p. 5—7 : 1 fig. 1 tab. 5 lit.
On the weights of the measurements when adjusting angle-distance networks (BUL)
AS 12.15.33/12.15.37
Key words: Adjustment, geodetic network

The weight determination in the adjustment of heterogeneous observations, i.e. by networks of measured both angles and distances, is discussed. It is demonstrated that a comparative estimation of the accuracy of angular and length measurements is not necessary. The participation of distances and angles in the same measuring units, in which their mean square errors are given, is sufficient.

Georgiev, National Correspondent, Bulgarian IGD 84/I 254

528.021.7:528.11

Rusjaeva, E. A.
Ob ispol'zovanii preobrazujushhejj funkcii pri issledovanii rjadov oshibok geodezicheskikh izmerenijj
Izvest. vyssh. uchebn. Zaved., Geod. i Aehrofotos"jomka. — Moskva (1982) 5. — p. 20—27 : 5 lit.
On utilization of the transformation function for the analysis of errors of geodetic measurements (RUS)
AS 12.15.33/12.15.37/73.07
Key words: Electrooptical distance meter, transformation of measuring

A transformation of an asymmetrical inhomogeneous (as to the group variances) series of absolute values of double measurement differences obtained by electrooptical distance measuring instruments is carried out. By using Pearson- and Barlett-tests the series of the transformed values is shown to be homogeneous and normally distributed. The further investigation of the transformed values may be substantiatedly continued by regression analysis methods.

Dobrokhotov, National Correspondent, USSR IGD 84/I 255

528.022/.023.62:528.526.6

Seredovich, V. A.
Ob ispol'zovanii giroteodolitov pri stvornykh izmerenijakh
Geodez. i Kartogr. — Moskva (1982) 9. — p. 24—27
On utilization of gyro-theodolites for alignment measurements (RUS)
AS 12.15.33/73.07
Key words: Alignment, gyro-theodolite
Dobrokhotov, National Correspondent, USSR IGD 84/I 256

528.021.6/.7(79)

Slater, L. E.
Multiwavelength EDM measurements in Southern California (ENG)
Proc. General Meet. IAG, Tokyo, May 7—15, 1982. — Tokyo (1982). — p. 268
AS 12.15.33
Key words: Electromagnetic distance measurement, geodetic network, multiwavelength distance measurement instrument, strain
Nagasawa, National Correspondent, Japan IGD 84/I 257

528.247:531.35

Tarakanov, Ju. A.
Vydelenie gravitacionnykh anomalijj pri izmerenijakh v dvizhenii
Dokl. AN SSSR. — (1982) 3. — p. 583—586
Determination of gravity anomalies from measurements in motion (RUS)
AS 12.15.33/12.15.09
Key words: Dynamics, gravity anomalies
Dobrokhotov, National Correspondent, USSR IGD 84/I 258

528.27(26)

Vatamanovskijj, Eh. A. / Izmajjlov, Ju. P. / Markov, G. S.
Morskaja gravimetricheskaja s"jomka v uslovijakh arkticheskikh morejj
Fiz.-tekhn. gravimetrija: apparatura i metody izmerenijj. — Moskva (1982). — p. 149—160 : 12 lit.
Marine gravimetric survey under conditions of Artic seas (RUS)
AS 12.15.33/12.15.09/12.15.21
Key words: Gravity measurement, marine geodesy
Dobrokhotov, National Correspondent, USSR IGD 84/I 259

528.35.02(437):551.244

Vyskocil, P.
Horizontal deformations on the border of the Bohemian Massif as determined by repeated EDM measurements (ENG)
Proc. General Meet. IAG, Tokyo, May 7—15, 1982. — Tokyo (1982). — p. 273
AS 12.15.33/12.15.09
Key words: Deformation measurement, electronic distance measurement, horizontal deformation, vertical measurement

In order to study the relative motions between the Bohemian Massif and the Carpathians the special geodetic networks of vertical as well as horizontal measurements were established during the 70's. The results of repeated distance measurements are transformed in the field of deformations and the main axis and measure of stress-strain deformations is determined. Using this kind of analysis the former results on the spreading tendencies between both geological systems are confirmed. The results on surface deformations are discussed considering the deep structure of the area under study, since the main deformations are connected with the area of strong subsidence of Bohemian Massif under the Carpathians. The results are compared with those obtained using geodetic as well as in situ measurements within the territory of Central Europe.

Nagasawa, National Correspondent, Japan IGD 84/I 260

530.152.1(71)

Webb, Jon D. / Penney, Robert C.
Six years of inertial surveying at geodetic survey of Canada (ENG)
Energy, Mines and Resources Canada, Surveys and Mapp. Branch. – Ottawa (1981). – p. 181–199 : 4 fig. 5 tab.
AS 12.15.33/12.15.25
Key words: Canada, inertial system

This paper will deal with geodetic survey's inertial survey operations from 1978 through 1980, with a brief outline of work done in previous years. The emphasis will be on improvements in the on-line software acquired by geodetic survey (namely Auto 23), improvements in the off-line field data processing and storage, measuring procedures, auxiliary equipment and logistics. Since costs are important they are dealt with as extensively as possible.

(Author's summ.)

Techn. Univ. Dresden IGD 84/I 261

530.152.1:528.283

Wilkinson, G. A. / Wong, R. V. C.
The use of azimuth control in inertial surveying to improve accuracy on an L-shaped traverse (ENG)
Energy, Mines and Resources Canada, Surveys and Mapp. Branch. – Ottawa (1981). – p. 219–231 : 9 fig. 3 tab. 4 lit.
AS 12.15.33/12.15.13/12.15.25
Key words: Azimuth measurement, inertial surveying system

Experience and controlled tests with the Litton inertial survey system have shown that inertial survey position accuracies are significantly degraded when a traverse departs from a straight line between control stations. Error propagation and simulation techniques that attempt to model the observed error behaviour indicated that a considerable improvement in the accuracy of an L-shaped traverse could be achieved if the initial azimuth misalignment was reduced either by providing external azimuth measurement or additional co-ordinate updates at certain points in the traverse.

(Author's summ., abbr.)

Techn. Univ. Dresden IGD 84/I 262

AIDOS Systematics **12.15.37** **Group I Geodesy**

Mathematical treatment of geodetic measurements

Theory and methods of the mathematical treatment of geodetic measurements
Theory of observation errors and method of least squares
Adjustment and catalogue of the results of measurement
Geodetic calculations
Modern techniques of mathematics and computation for geodetic calculations

Gruppe I Höhere Geodäsie

Mathematische Bearbeitung geodätischer Messungen

Theorie und Methoden der mathematischen Bearbeitung geodätischer Messungen
Theorie der Meßfehler und Methode der kleinsten Quadrate
Ausgleichung und Katalogisierung der Meßergebnisse
Geodätische Berechnungen
Moderne Verfahren der Mathematik und Rechentechnik für geodätische Berechnungen

528.181

Baetsle, P.-L.
Encore le calcul matriciel (suite)
Les moindres carres
Bull. trimestr. Soc. belge Photogramm., Ann. D 10, D 11, D 12. — Bruxelles (1981—1983) 145—150. — p. 135—183
On matrix calculus (sequel)
Least squares (FRE)
AS 12.15.37
Key words: Matrix calculus, least squares method

Techn. Univ. Dresden IGD 84/I 263

528.3/.4 :528.14 :519.2

Carosio, Allessandro
Multivariate Statistik und Zuverlaessigkeit
Vermess. Photogramm. Kulturtechnik. — Zuerich 81 (1983) 8. — p. 265—274 : 8 fig. 20 lit.
Multivariate statistics and reliability (GER)
AS 12.15.37/12.15.25
Key words: Adjustment, geodetic network, mathematical geodesy, mathematical statistics

Geodesists have always been interested in two important problems: planning and development of reliable measuring arrangements with appropriate calculation methods, and the determination and localization of gross errors after the measurements. The present article is an account of the two problems and demonstrates the relations between statistical tests and reliability as well as between intuitive techniques of practice and modern methods of multivariate statistics. Description of a new test for the evaluation of triangulation results (NMAX test) and its effect on the reliability of measuring systems is demonstrated.

Techn. Univ. Dresden IGD 84/I 264

528.02 :621.396.962
528.28 :629.783

Herzberger, Kurt
Modifizierung des Programmsystems GEODOP zur Nutzung verschiedenartiger Dopplerdaten
Veroeff. Bayer. Kommiss. internat. Erdmess., astron.-geod. Arb. — Muenchen (1981) 41. — p. 69—76 : 2 fig. 1 tab. 7 lit.
Modification of the programme system GEODOP for utilisation of different Doppler data (GER)
AS 12.15.37/12.15.17
Key words: Doppler measurement, satellite

The programme system GEODOP serves to determine the position of one or more observation stations from Doppler observations to the satellites of the U.S. Navy Navigation Satellite System (NNSS/TRANSIT).

The use of newly developed Doppler receivers, some of which put out different information and also widely, differing data formats, requires continual adaptation of the programmes. For the evaluation work carried out by the SFB 78 as part of international Doppler observation campaigns it was vitally necessary to continually adapt the software to the wide variety of data output. It is the aim of this report to describe the modifications and to put forward a suggestion for a uniform data format for reciprocal data exchange between the numerous users of the GEODOP programme.

Schoen, National Correspondent, Federal Republic of Germany IGD 84/I 265

528.14:528.42
528.42:528.14

Kougija, V. A.
Ob adekvatnosti metoda naimen'shikh kvadratov additivnomu predstavleniju o strukture oshibok izmerenijj
Markshejjd. delo i geod. — Leningrad (1982). — p. 11—13
On adequacy of the least-squares method to the additive concept about the structure of measuring errors (RUS)
AS 12.15.37/12.15.33
Key words: Least-squares method, measuring error

The separation of measuring errors into any number of components and the adjustment under condition of the minimum of the sum of corrections squared for these components are shown to be adequate to the least-squares solution without partition into components. Therefore, the least-squares method is in agreement with the additive concept about the structure of measuring errors.

Dobrokhotov, National Correspondent, USSR IGD 84/I 266

528.112

LaRiva, Camilo de / Junker, Armin
Der „systematische Fehler“ — kein schluessiger Begriff fuer die Kennzeichnung von Messverfahren
DIN-Mitt. — Berlin 61 (1982) 6. — p. 339—341 : 5 lit.
"Systematic error" — not a conclusive term for describing measuring procedures (GER)
AS 12.15.37/12.15.33
Key words: Systematic error

The quality of the transformation of the information obtained during calibration with respect to the defined portion of the relation "primary measuring signal/condition" is decisive for the occurence and magnitude of "systematic errors". The fact that conditions of objects measured in the calibration experiment can only be simulated approximately by corresponding reference material was taken into account.

Therefore systematic errors can be avoided on principle, i.e. they are independent of the type of measuring procedure.

Thus, the term "systematic error" is not suited to objective classification of measuring procedures; it simply serves to describe the respective standards used by the agency developing the measuring procedure for the transformation procedure "primary measuring signal/condition". Hence, the systematic error is of subjective character and the corresponding term can be standardized to a limited extent only.

Schoen, National Correspondent, Federal Republic of Germany IGD 84/I 267

528.23:681.3

Lazarev, V. M.
Primenenie rekurrentnykh formul dlja reshenija geodezicheskikh zadach na EhVM
ONTI CNGIIGAiK. — (1982) Nr. 94gd-DEP82. — 14 p. : 3 lit.
Use of recursive formulas for a computer-based solution of geodetic problems (RUS)
AS 12.15.37/63.23
Key words: Gaussian co-ordinate system, recursive formula

Algorithms and descriptions of the programs written on the language Fortran-IV are presented. The programs are intended for solution of the principal geodetic problem in a Gaussian co-ordinate system. The recursive formulas used in the programs allow computations of rectangular and geodetic co-ordinates of the stations situated within 120°- and 60°-zones with errors less than 1 mm and 0.0001″.

A special program provides for solution of normal equation systems by the method of comparison of coefficients that makes it possible to determine both the unknowns and their weights.

Dobrokhotov, National Correspondent, USSR IGD 84/I 268

517.5:519.281

Lilienblum, Horst
Parameterschaetzungen von Zeitreihen
Vermessungstechnik. — Berlin 32 (1984) 1. — p. 24—26 : 5 fig. 1 tab. 5 lit.
Parameter estimation of time series (GER)
AS 12.15.37
Key words: Time series

The determination of empirically obtained estimates of mathematical-statistical parameters of time series is performed and illustrated by a numerical example and some results and their accuracies are discussed. Several graphical representations demonstrate the essential connection and typical properties of the process.

(Author's summ.)

Techn. Univ. Dresden, IGD 84/I 269

528.385-187.4:519.233

Lindner, Bernd-Ulrich
Mathematische Modelle und Hypothesentests bei der Bestimmung vertikaler Punktbewegungen durch Praezisionsnivellements
Diss. TU Berlin, Fachber. Bauing.- u. Vermess.-Wesen
Dt. geod. Kommiss., R. C. — Muenchen (1983) 280. — 104 p. : 14 fig. 14 tab. 109 lit. 6 app.
Mathematical models and hypothesis test in the determination of vertical point movements by precision levelling (GER)
AS 12.15.37/12.15.05
Key words: Hypothesis test, mathematical models of precision levelling, parameter estimation, point movements

Deriving from a general linear model of levelling (which considers different random and systematic errors as well as point movements during measurement) simpler models are investigated with respect to fundamental model hypotheses. For the estimation of model parameters and variance-covariance components different estimation techniques are described on the basis of a variance-covariance model.

It is shown that different models require different test techniques for localization of unstable points. If the standard model is extended by surveys (and we are not able to check them) a high uncertainty of results appears due to model errors.

Techn. Univ. Dresden IGD 84/I 270

341.22(493):681.3.07

Loon, C. van
Opstelling en gebruik van een gegevensbank van de administratieve grenzen van Belgie
Bull. trimestr. Soc. belge Photogramm. et de Teledetec. — Bruxelles (1983) 149—150. — p. 3—8
Creation and use of a data bank of the Belgian administrative boundaries (DUT)
AS 12.15.37/63.23
Key words: Administrative boundary, data bank

Nowadays interactive terminals are one of the chief elements of systems designed for numerical cartography. They allow to work out data banks, to update them and to make use of them. This article deals with the use of an interactive system at the National Geographical Institute with a view to working out a data bank of the administrative boundaries. It also describes the way these data can be applied.

(Author's summ.)

Techn. Univ. Dresden IGD 84/I 271

528.33:528.14

Reichsthaler, K.
Dreidimensionale Netzausgleichung im Testnetz Steiermark unter Beruecksichtigung von Dopplermessungen
Mitt. geod. Inst. TU Graz. — Graz (1983) 46. — 152 p.
Three-dimensional net adjustment in the testnet Styria under consideration of Doppler measurements (GER)
AS 12.15.37/12.15.33
Key words: Adjustment, Doppler measurement, three-dimensional network

In this paper the theory of the adjustment of a three-dimensional network is described. The computer software published by the same author in a previous paper is extended to include also Doppler coordinates and applied to the national survey. In particular, the testnet Styria was examined. Considering the inner accuracies it was shown that Doppler coordinates do not improve point accuracies as determined from terrestrial measurements, at least for small areas. However, Doppler-observed heights agree with ellipsoidal heights satisfactorily and the influence in the reduction to ellipsoidal coordinates.

(Author's summ.)

Erker, National Correspondent, Austria IGD 84/I 272

528.113

Reissmann, G.
Was ist der Unterschied zwischen dem mittleren Fehler und der Standardabweichung?
Vermessungstechnik. — Berlin 31 (1983) 11. — p. 382—383
What is the difference between the mean error and the standard deviation? (GER)
AS 12.15.37
Key words: Definition, mean error, standard deviation

Explanations of origin and definition of the accuracy measures "mean error" and "standard deviation". Considerations of their common properties and differences and suggestions for their common and future application.

(Author's summ.)

Techn. Univ. Dresden IGD 84/I 273

528.142:528.331(430.1)

Schmidt, Rudolf
Zur Diagnoseausgleichung des Deutschen Hauptdreiecksnetzes.
Problemstellung
Dt. geod. Kommiss., R. B. — Frankfurt/M. (1982) 262. — p. 11—43 : div. lit.
1 app. (= Mitt. Nr. 165 Inst. Angew. Geod.)
Diagnostic adjustment of the German primary triangulation net.
Formulation of problem (GER)
AS 12.15.37/12.15.25
Key words: Diagnostic adjustment, triangulation net

The marginal parts of the net were, wherever possible, placed so far away from the border of the Federal Republic of Germany that instabilities caused by marginal effects are not to be feared within the interior of the country. The adjustments by observation equations in the system of geographical co-ordinates were carried out by IfAG on the computer TR 440 by means of a computing program developed by IfAG.

To investigate the accuracy of the neighbouring points, further computing operations were carried out such as geodetic connections of small net parts or transformations with terms of higher order and modern procedures applied, i.e. procedures which are based on collocation and permit a separation of the errors into trend, signal and noise.
Schoen, National Correspondent, Federal Republic of Germany IGD 84/I 274

528.02.06:519.2

Steeves, Robin R.
A statistical test for significance of peaks in the least squares spectrum (ENG)
Energy, Mines and Resources Canada, Surveys and Mapp. Branch. — Ottawa (1981). — p. 149—165 : 7 lit.
AS 12.15.37/12.15.05
Key words: Mathematical statistics, spectral analysis, statistical test

After a review of the method of least squares spectral analysis (Vanicek, 1971), with the inclusion of an arbitrary non-singular covariance matrix for the observed time series, a statistical test is developed for testing the significance of peaks in the least squares spectrum. A by-product of this investigation is the derivation of the probability density function of the least squares spectrum under the hypothesis that the observed time series is derived from a vector of statistically independent random variables whose probability density function is the multidimensional normal distribution $N_n(0, \sigma^2 1)$. A simple formula is given for computing a $100(1-\alpha)\,\%$ critical value for significant peaks in the spectrum. (Author's summ.)
Techn. Univ. Dresden IGD 84/I 275

528.1.001.5

Wolf, Helmut
Neues Altes in der Ausgleichungsrechnung
Vermess. Photogramm. Kulturtechnik. — Zuerich 81 (1983) 7. — p. 233—240 : 70 lit.
New old findings in adjustment calculus (GER)
AS 12.15.37
Key words: Adjustment

New models and techniques for adjustment calculus developed in recent times are investigated with respect to their common features with classical methods. It is described that the most recent ideas show a close connection with the ideas of Gauss and Helmert. In this respect some points are considered in detail: adjustment principles, observation equations, correlated observations, prediction and collocation, free adjustments, co-ordinate transformations on the basis of net adjustments and the latest findings concerning the Gauss-Helmert model of adjustment calculus.
Techn. Univ. Dresden IGD 84/I 276

AIDOS Systematics **73** **Group I Geodesy**

Design, manufacture and proof of instruments

Annotation: Problems of theory, calculation, designing, manufacture, examination and proof of instruments and devices (for applications see the respective sections)

Gruppe I Höhere Geodäsie

Konstruktion, Fertigung und Prüfung von Geräten

Anmerkung: Fragen der Theorie, Berechnung, Projektierung, Fertigung, Untersuchung und Prüfung von Geräten (Anwendungen der Geräte in den entsprechenden Abschnitten)

AIDOS Systematics **73.07** **Group I Geodesy**

Construction and proof of geodetic instruments

Distance measuring instruments
Instruments for measuring angles and directions
Instruments for measuring height and inclination
Plotting tables, combined devices
Instruments for measuring oriented directions
Gyroscopes, compasses
Accessories

Gruppe I Höhere Geodäsie

Konstruktion und Prüfung geodätischer Geräte

Geräte zur Streckenmessung
Geräte zur Winkel- und Richtungsmessung
Geräte zur Höhen- und Neigungsmessung
Kartiertische, Gerätekombinationen
Geräte zur Messung orientierter Richtungen
Kreisel, Bussolen
Zusatzgeräte

528.517.8.089.6:629.783

Etling, Walter et al.
Kalibrierungsmessungen mit dem Nd:YAG-Lasermesssystem zur Bestimmung der Additionskonstanten
Veroeff. Bayer. Kommiss. internat. Erdmess. — Muenchen (1981) astron.-geod. Arb. 41. — p. 53—68 : 9 fig. 3 tab. 5 lit.
Calibration measurements using the Nd:YAG laser measurement system to determine additive constants (GER)
AS 73.07/73.19
Key words: Calibration, laser distance measurement

Distances determined on the basis of transit times normally have to be centered on the geometric reference point of the measurement system. In general this is taken to be the intersection of azimuth and vertical axes. Since the beginning of satellite ranging at the Wettzell station at least one calibration measurement is taken for every successful orbit determination in order to ascertain the necessary amount of correction. This is determined by comparing a distance measured e.g. by means of a geodimeter with the result of the transit time measurements of the laser pulses. Instrumental arrangement, performance and results of test measurements are described.
(Author's summ.)

Schoen, National Correspondent, Federal Republic of Germany IGD 84/I 277

532.291.089.62-187.4

Geiger, Axel
Entwicklung und Erprobung eines Praezisionsneigungstisches zur Kalibrierung geodaetischer Instrumente
Hochsch. Bundeswehr Muenchen, wiss. Stud.-Gang Vermess.-Wesen, Schr.-R. — Muenchen (1982) 8. — 124 p. : 55 fig. 15 tab. 72 lit.
Development and test of an inclination table of high precision for calibration of geodetic instruments (GER)
AS 73.07
Key words: Calibration, geodetic instrument, level tester

Considerations and reasons for development and construction of an inclination table are described in detail. The inclination table mainly serves for the calibration of levels and compensators. The inclination ranges from approx. ± 2 gon to ± 4 gon. The measuring principle is based on the interferometric measurement of differences in distance by a laser interferometer. The accuracy of an inclination measurement amounts to ± 0.01 mgon for an inclination angle of ≦ 1 gon. In the second part of this work applications of the inclination table are described. In this way, a geodetic test instrument of high reliability and precision is available.

Techn. Univ. Dresden IGD 84/I 278

528.52/.53(494):528.4

...
Informatik-Theodolit Theomat Wild T 2000
Vermess. Photogramm. Kulturtechnik. — Zuerich 81 (1983) 7. — p. 254—256 : 3 fig.
Theodolite Theomat Wild T 2000 (GER)
AS 73.07
Key words: Theodolite T 2000

Techn. Univ. Dresden IGD 84/I 279

528.517.021.7.088

Scherer, Michael
Entwicklung eines weitreichenden Praezisionsdistanzmessers als Grundlage fuer ein Mehrwellengeraet — Habilitation Rhein. Friedr.-Wilhelms-Univ. Bonn
Dt. geod. Kommiss., R. C. — Muenchen (1982) 272. — 108 p. : div. fig. div. tab. div. lit.
Development of a high precision range finder for long distances as the basis for a multi-wave instrument — Habilitation (GER)
AS 73.07
Key words: Electrooptical range finder

The prototype of a high precision electrooptical range finder for measuring distances up to 30 km is described in detail.

The measuring system consists of a mode stabilized HeNe-lasertube which emitts light varying its type of polarization at 640 MHz. On its way to the detecting photodiode a small part of this light-beam is linearly polarized while the mean part is directed to a cube corner. The reflected light-beam passes through a rotating linear polarizer before it is mixed in the diode with the beam from the inner path. The amplitude of the resulting output voltage is a function of the angular position of the rotating filter and of the phase delay between the signals of the internal and external light pathes. The angular position of the rotating filter and the sinuosidal output voltage of the diode are continuously correlated using phase-locked loop technique. Results better than ± (0.2 mm + 0.2 ppm) have been achieved.

Schoen, National Correspondent, Federal Republic of Germany IGD 84/I 280

528.521.087.7:681.3

Shul'c, V. G. / Svetlov, V. P. / Brejjdo, I. I.
Kodovyjj teodolit TT II
Geod. i Kartogr. — Moskva (1982) 9. — p. 31—34
A code theodolite TT II (RUS)
AS 73.07
Key words: Code theodolite

Dobrokhotov, National Correspondent, USSR IGD 84/I 281

528.541.82-187.4:528.089.6(71)

Vamosi, S.
Calibration procedures for invar levelling rods in Canada (ENG)
Energy, Mines and Resources Canada, Surveys and Mapp. Branch. – Ottawa (1980). – p. 83–97 : 7 fig. 1 tab. 9 lit.
AS 73.07/12.15.33
Key words: Calibration, levelling rod

The first use of the field calibrator portable type LK (Quartz Ceramic) is described and a summary of calibration results is presented. Calibration results obtained with the field calibrator portable type have been compared with those of the HP5526A laser measurement system. From the comparison one may conclude that the field calibrator portable type LK is an excellent standard and could provide sufficiently precise checks on the invar rods during field operations.

(Author's summ.)

Techn. Univ. Dresden IGD 84/I 282

528.517.089

Vershinin, V. I.
K metodike opredelenija postojannykh popravok svetodal'nomerov
Geod. i Kartogr. – Moskva (1982) 9. – p. 9–12
The methods of determination of constant corrections of electrooptical distance measuring instruments (RUS)
AS 73.07
Key words: Electrooptical distance meter

Dobrokhotov, National Correspondent, USSR IGD 84/I 283

528.517.089.6

Wilkinson, G. A.
Report on Mekometer calibrations during the 1980 field season (ENG)
Energy, Mines and Resources Canada, Surveys and Mapp. Branch. – Ottawa (1981). – p. 201–218 : 1 fig. 7 tab.
AS 73.07
Key words: Calibration, Mekometer

A brief account is given of initial testing of the Kern ME3000 Mekometer on the national geodetic baseline to determine its accuracy and the nature of its instrumental errors, observation procedures for a satisfactory determination of instrument calibration parameters and the development of software to handle the data.

The results of calibration measurements taken during 1980 are closely examined and interpreted to form a basis on which to select appropriate parameter values to reduce measurements made on various calibration baselines.

(Author's summ., abbr.)

Techn. Univ. Dresden IGD 84/I 284

528.531

Zakharov, A. I.
Nomogrammnyjj takheometr 2TN
Geod. i Kartogr. – Moskva (1982) 11. – p. 53–55
The nomogram tacheometer 2TN (RUS)
AS 73.07
Key words: Tacheometer

Dobrokhotov, National Correspondent, USSR IGD 84/I 285

AIDOS Systematics **73.19** **Group I Geodesy**

Geophysical instruments

— for meteorology and oceanography
— for determining the gravity field and the magnetic field
— for seismology

Gruppe I Höhere Geodäsie

Geophysikalische Geräte

— für Meteorologie und Ozeanographie
— für die Bestimmung des Schwere- und Magnetfeldes
— für die Seismik

528.563.089.6

Zheleznjak, L. K. / Ehlinson, L. S.
Osobennosti ehtalonirovanija gravimetra s dvumja uprugimi sistemami krutil'nogo tipa metodom naklona
Fiz.-tekhn. gravimetrija: apparatura i metody izmerenijj. — Moskva (1982). — p. 110—124 : 4 lit.
Peculiarities of calibrating a gravity-meter with two elastic systems of torsional type by the tilting method (RUS)
AS 73.19
Key words: Calibration, gravimeter
Dobrokhotov, National Correspondent, USSR IGD 84/I 286

528.563(26)

Zheleznjak, L. K. / Popov, E. I.
Principy postroenija i optimal'naja skhema sovremennogo morskogo gravimetra
Fiz.-tekhn. gravimetrija: apparatura i metody izmerenijj. — Moskva (1982). — p. 43—60
The principles of construction and the optimal scheme of an up-to-date marine gravity meter (RUS)
AS 73.19/12.15.21
Key words: Gravimeter, marine geodesy
Dobrokhotov, National Correspondent, USSR IGD 84/I 287

AIDOS Systematics **12.23** **GROUP II SURVEYING AND TOPOGRAPHY**

GRUPPE II ANGEWANDTE GEODÄSIE UND TOPOGRAPHIE

AIDOS Systematics **12.23.05** **Group II Surveying and topography**

Geodetic foundations of topographic and engineering-geodetic work

Establishing of local horizontal and vertical control networks, construction and factory networks
Filling-in of first-order networks for engineering-geodetic tasks
Geodetic determination of tie points and control points
Measurements methods (intersection methods, traversing, detailed triangulation) and evaluation methods for establishing local control networks

Gruppe II Angewandte Geodäsie und Topographie

Geodätische Grundlagen für topographische und ingenieurgeodätische Arbeiten

Anlage von lokalen Lage- und Höhenfestpunktnetzen, Bau- und Werksnetzen
Verdichtung von Landesnetzen für ingenieur-geodätische Aufgaben
Geodätische Bestimmung von Anschluß- und Paßpunkten
Messungsmethoden (Einschneideverfahren, Polygonometrie, Kleintriangulation) und Auswerteverfahren zur Anlage lokaler Festpunktnetze

528.335.088:528.42

Alejjnikov, S. A.
Voprosy tochnosti s"jomochnogo obosnovanija topograficheskikh s"jomok v masshtabe 1 : 500
Geod. i Kartogr. — Moskva (1982) 11. — p. 32—35 : 4 lit.
The accuracy of geodetic control for topographic surveys at the scale 1 : 500 (RUS)
AS 12.23.05/12.23.13
Key words: Geodetic control net, topography

Dobrokhotov, National Correspondent, USSR IGD 84/II 46

528.48

Bagratuni, G. V. et al.
● **Spravochnik po geodezicheskim razbivochnym rabotam**
Moskva: Nedra. 1982. 129 p. : 8 lit.
A reference book for geodetic setting-out operations (RUS)
AS 12.23.05
Key words: Engineering surveying

Dobrokhotov, National Correspondent, USSR IGD 84/II 47

528.414

Basova, I. A.
O tochnosti postroenija prostranstvennojj poligonometrii
Geod. i fotogramm. — Rostov n/D. (1981). — p. 46—54 : 9 lit.
On the accuracy of establishing three-dimensional traverses (RUS)
AS 12.23.05/12.15.33
Key words: Traversing
Dobrokhotov, National Correspondent, USSR IGD 84/II 48

528.411/.412

Brandstaetter, Gerhard
Die analytische Loesung der simultanen Doppel- und Mehrfachpunktbestimmung in der Ebene
Oesterr. Z. Vermess.-Wesen u. Photogr. — Wien 71 (1983) 2. — p. 65—71
The analytical solution of a simultaneous double and multiple point determination in the plane (GER)
AS 12.23.05
Key words: Critical configurations, directions, simultaneous point determination

The problem of the non-redundant simultaneous determination of two and more unknown points by means of measured directions at these points to their neighbours and to given control points is solved analytically. This leads to a system of 2 n + 1 linear equations with 2 n unknown coordinates and the unknown orientation of the chain. The solution for the latter enables a discussion of the numerical influence of critical configurations.
(Author's summ.)
Erker, National Correspondent, Austria IGD 84/II 49

528.411:528.58

Cenev, St. N.
Edin metod za opredeljane na izkhodnata velichina pri izvahrshhvane orazmerjavaneto na konstrukcii za stabilizirane na geodezicheskite tochki
Geod. Kartogr. Zemeustr. — Sofia (1983) 1. — p. 12—15
A method of initial value determination when dimensioning structures of geodetic points stabilization (BUL)
AS 12.23.05
Key words: Control point determination
Georgiev, National Correspondent, Bulgaria IGD 84/II 50

528.411:528.11.023.088

Korobkov, S. A. / Orlov, Ju. I.
Ocenka tochnosti pri uravnivanii planovykh setejj v prirashhenijakh koordinat
Markshejjd. delo i geod. — Leningrad (1982). — p. 30—34 : 2 lit.
Accuracy estimation of co-ordinate differences in adjustment of horizontal control nets (RUS)
AS 12.23.05/12.15.37
Key words: Adjustment, geodetic network
Dobrokhotov, National Correspondent, USSR IGD 84/II 51

528.024:528.118

Kostadinov, K.
Ocenka na tochnostta na vkljuchen nivelachen khod
Geod., Kartogr., Zemeustrojjstvo. — Sofija (1983) 2. — p. 8—11 : 4 fig. 1 tab. 6 lit.
Accuracy estimation of connected levelling line (BUL)
AS 12.23.05/12.15.37
Key words: Accuracy estimation, adjustment, levelling network

The accuracy estimation of heights of intermediate bench marks by adjusting a line of geometric or trigonometric levelling between two given points is examined. Respective formulae are obtained and discussed and their connection to some earlier results published by V. I. Sal'nikov in the USSR is indicated. An example is given and the possibilities of utilization of the formulae in the automated adjustment of levelling networks is discussed.

Georgiev, National Correspondent, Bulgaria IGD 84/II 52

528.33.088:528.45

Losev, K. A.
Prognozirovanie nadjozhnosti geodezicheskogo obosnovanija na zastroennykh territorijakh
Geod. kontrol' v str-ve. — Kujjbyshev (1982). — p. 79—84
Prediction of reliability of geodetic control in the built-up areas (RUS)
AS 12.23.05/12.23.09
Key words: Geodetic network, prediction

Dobrokhotov, National Correspondent, USSR IGD 84/II 53

528.412.088.3

Pauli, Wolfgang
Zur Genauigkeit der Standpunktbestimmung
Vermessungstechnik. — Berlin 32 (1984) 1. — p. 22—23 : 2 tab. 3 lit.
The accuracy of position determination (GER)
AS 12.23.05
Key words: Position determination

By the application of the law of error propagation the formulae of the mean positional errors are determined for the following four techniques and are compared by means of examples: combined resection/arc section (RE/BS) without determination of the scale factor; corresponding to this, with scale factor from constants for a Helmert transformation (HT); arc section (BS); resection (RE).

(Author's summ.)

Techn. Univ. Dresden IGD 84/II 54

528.412.5.088

Skazkin, I. I.
O tochnosti prjamojj uglovojj prostranstvennojj zasechki
Inzh.-geod. raboty pri str-ve, montazhe i ehkspluat. inzh. sooruzh. — Moskva (1982). — p. 50—57
On the accuracy of the angular three-dimensional intersection (RUS)
AS 12.23.05/12.15.33
Key words: Angular intersection

Dobrokhotov, National Correspondent, USSR IGD 84/II 55

528.58

Ulrich, W.
Die Verwendung von Kunststoffgrenzzeichen in der Parzellarvermessung
Vermess. Photogramm. Kulturtechnik. — Zuerich 81 (1983) 8. — p. 274—279 : 5 fig.
The use of boundary marks of synthetic material in the surveying of land (GER)
AS 12.23.05/12.23.09
Key words: Boundary markers

Techn. Univ. Dresden IGD 84/II 56

528.411:528.11

Vajjnauskas, V. V.
Primenenie topologicheskikh metodov dlja optimizacii i uravnivanija inzhenerno-geodezicheskikh setejj
Tr. po geod. Vil'njus. inzh.-stroit. in-t. – Vilnjus (1982) 11. – p. 25–33 : 4 lit.
The use of topologic methods for optimization and adjustment of engineering geodetic nets (RUS)
AS 12.23.05/12.15.37
Key words: Adjustment, engineering geodesy, topology

By numerical examples, the practical side of utilization of topological methods for adjustment and optimization of geodetic measurements is illustrated. It is concluded that the method is more efficient than statistical or covariance methods that involve complicated matrix transformations.

Dobrokhotov, National Correspondent, USSR IGD 84/II 57

528.425.4:519.246

Wild, Emil
Die Praediktion mit Gewichtsfunktionen und deren Anwendung zur Beschreibung von Gelaendeflaechen bei topographischen Gelaendeaufnahmen – Diss.
Dt. geod. Kommiss., R. C. – Muenchen (1983) 277. – 139 p. : div. fig. div. tab. 61 lit.
The prediction of weight functions and its application to the description of terrain areas in topographic surveying – Diss. (GER)
AS 12.23.05/12.15.37
Key words: Prediction, topographic surveying

Since the end of the fifties a variety of techniques and programmes were developed for the automatic evaluation of relief surveys and for the creation of digital relief models. The most important techniques are described in the first two chapters. They are analyzed by means of weight functions. In the last two chapters a technique of prediction with weight functions is developed. Efficiency and suitability of different weight functions are studied. Description of solutions for surface smoothing for the purpose of minimization of measuring errors and roughness as well as for consideration of marked points and lines. By means of practical and simulated examples, various applications are shown.

Techn. Univ. Dresden IGD 84/II 58

AIDOS Systematics **12.23.09** **Group II Surveying and topography**

Topographic surveying on land

Obtaining information about the terrain by using geodetic methods (tacheometric surveying, plane table surveying and other methods) and means for field survey and for supplementary measurements
Cadastral surveying

Gruppe II Angewandte Geodäsie und Topographie

Topographische Aufnahme des Festlandes

Gewinnung von Informationen über das Gelände mit geodätischen Methoden (tachymetrische Aufnahme, Meßtischaufnahme und andere Verfahren) und Mittel zur Geländeaufnahme und für Ergänzungsmessungen
Katastervermessung

528.44:331.875

Beek, G. S. C. van
Hoe staat het met AKR?
Geodesia Nederl. geod. T. — Apeldoorn 24 (1982) 1. — p. 5–9 : 11 fig.
Status of automation of cadastral registration (DUT)
AS 12.23.09
Key words: Cadastre, land information system

Automation of Cadastral Registration (ACR) and its consequences for personnel and organization are elaborated upon. ACR will serve as a basis for (cadastral) land information systems. Prevailing views on the importance of ACR tend to slow down the development of these systems. Operational data bases as well as implementation of ACR in the province of Zeeland are described.
(Author's summ.)
Pouls, National Correspondent of the Netherlands IGD 84/II 59

528.44

Claessen, W. A.
De rol van het Kadaster bij de vastgoedinformatie
Geodesia Nederl. geod. T. — Apeldoorn 24 (1982) 1. — p. 2–4 : 4 fig. 5 lit.
The role of the cadastral service in developing land information systems (DUT)
AS 12.23.09
Key words: Cadastre, land information system

After describing the contents and quantity of present-day cadastral registers and maps the importance of the cadastre's role in developing land information systems is stressed upon. Prevailing questions and problems around this role are summed up. Guidelines to improve status of land information systems within the government are given.
(Author's summ.)
Pouls, National Correspondent of the Netherlands IGD 84/II 60

528.425.2:528.531

Erkelens, J.
Tachymetrie bij de Grontmij
Geodesia Nederl. geod. T. — Apeldoorn 24 (1982) 7/8. — p. 256–262 : 8 fig.
Tacheometry by the Grontmij (DUT)
AS 12.23.09
Key words: Tacheometer, tacheometry

A brief description is given how, since 1974, tacheometric measurements are coded and processed using a Harris Computer and a Calcomp drumplotter. The coding is based on a line numbering system. Since 1981 the "Grontmij" uses the GEOS tacheometric system of Zeiss, which is based on a point numbering system. The complete procedure from measurement to map and digital base is described.
(Author's summ.)
Pouls, National Correspondent of the Netherlands IGD 84/II 61

528.44

Gerstbach, G.
Absolute Orientierung in der Landes- und Katastervermessung
Oesterr. Z. Vermess.-Wesen u. Photogramm. — Wien 71 (1983) 2. — p. 72–81
Absolute orientation methods used in land- and cadastral surveying (GER)
AS 12.23.09/12.23.05
Key words: Polaris-azimuths, solar orientation, traversing

Solar orientation methods prove advantageous in cadastral surveying when no control points or connecting directions are available. There is no need for special equipment and accuracies of $\pm 10''$ ($\pm$ 3 mgon) may be obtained within a few minutes. For this, a supplement to the Austrian decree of cadastral surveying is proposed, that would suit for gyroscope observations too.

In geodetic networks Polaris-azimuths are preferable. Common theodolites can secure $\pm 1''$ and the observation might be performed even in daylight. Some examples show how effectively absolute orientations increase the strength of geodetic networks.
(Author's summ.)
Erker, National Correspondent, Austria IGD 84/II 62

528.422

Hofmann-Wellenhof, B.
Aussparungsflaechen und unregelmaessige Raender bei der Ableitung von Hoehenlinien aus einem Hoehenraster
Oesterr. Z. Vermess.-Wesen u. Photogr. — Wien 71 (1983) 2. — p. 83–94
Contour-free regions and irregular boundaries in the derivation of contour-lines from a height grid (GER)
AS 12.23.09
Key words: Derivation of contour-lines

A method for the treatment of contour-free regions and irregular boundaries in the derivation of contour lines from heights at a regular grid is proposed. The contour-free regions and the irregular boundaries, defined as closed polygons, are combined with the height-grid using linear lists. Therefore a limit does not exist neither for the number of point-coordinates of the contour-free regions nor for the number of polygons within a grid-mesh.
(Author's summ.)
Erker, National Correspondent, Austria IGD 84/II 63

528.44(436)

Kloiber, Otto
100 Jahre Evidenzhaltung des Grundkatasters
Oesterr. Z. Vermess.-Wesen u. Photogramm. — Wien 71 (1983) 2. — p. 57–63
100th anniversary of cadastral revision in Austria (GER)
AS 12.23.09
Key words: Cadastral revision, juridical background, organization
Erker, National Correspondent, Austria IGD 84/II 64

528.44(084.3)(492)

Koen, L. A. / Polman, J.
Perspectieven voor de kaartvervaardiging bij het Kadaster
Geodesia Nederl. geod. T. — Apeldoorn 24 (1982) 1. — p. 9–17
Prospects of mapmaking in the cadastral service (DUT)
AS 12.23.09/12.23.13
Key words: Cadastre, large scale map

A recently acquired task of the Cadastral Service (in the Netherlands) is the production, updating and publication of the Large Scale Basic Map of the Netherlands. Administrative and technical aspects of this task as well as implications for current cadastral mapping procedures are described. Subjects of discussion are the function of the map and quality requirements, remeasurements in the field and map revision, relation between the Large Scale Basic Map and cadastral maps, automation and financial-economic aspects.
Pouls, National Correspondent of the Netherlands IGD 84/II 65

528.44:528.412

Langenhan, Martin
Liegenschaftsvermessungen mit freier Standpunktwahl
Vermessungstechnik. — Berlin 32 (1984) 1. — p. 5–7 : 4 fig. 2 lit.
Cadastral surveying with free choice of stations (GER)
AS 12.23.09
Key words: Cadastral surveying, polar surveying

This rational method of determining stations during the performance of cadastral surveyings using polar surveying methods is described with the help of three examples. Therein the different geodetic pre-conditions are considered and the resulting necessary preparatory and evaluation works by means of electronic computer technology are described. Some possible analytical solution variants in the determination of station are mentioned.

(Author's summ.)

Techn. Univ. Dresden IGD 84/II 66

528.42:331.875
528.5-52:681.3

Mentes, Gyula
Problems of field data collection in geodetic measurements (ENG)
Acta Geod., Geoph. et Montan. — Budapest 18 (1983) 1—2. — p. 67—70 : 1 fig. 3 lit.
AS 12.23.09/73.07
Key words: Digital instrument, field data, geodetic instrument, surveying instruments

The paper deals with problems of the development of data collection systems for geodetic field measurements. The advantages of the semi-automatic and automatic data collection equipments are described in detail together with the most important demands on such systems. Finally, the construction of a new intelligent data collection system is described which eliminates the disadvantages of earlier data collection systems.

Veroe-Hetenyi, National Correspondent, Hungary IGD 84/II 67

528.4(492)(091)

Pouls, H. C.
Historisch onderzoek naar het werk van de landmeter
Futura. Inf. en Comm. blad van het Ned. Architecten Gezelschap. —
Den Haag 18 (1983) 2. — p. 17—23 : 10 fig. 7 lit.
Research into the history of landsurveying (DUT)
AS 12.23.09
Key words: Landsurveying, the Netherlands

Starting with the remark that in the Netherlands very little research into the history of landsurveying has been done by surveyors but mostly by others, the author states that errors have been made because of the fact that the writer knows nothing about surveying or surveying instruments. A number of examples are given, mostly in publications about old maps.

For a correct interpretation of survey in the past knowledge of old survey-instruments and survey methods is necessary.

When doing research work more attention should be paid to information about instruments used by surveyors in the past and the correct interpretation of the methods used and their possibilities.

(Author's summ.)

Pouls, National Correspondent of the Netherlands IGD 84/II 68

528.44:528.42

Raamsdonk, J. van
De verwerking van terestrische metingen met Systeem Detailmeting '76
Geodesia Nederl. geod. T. — Apeldoorn 24 (1982) 10. — p. 354—360 : 14 fig. 3 lit.
Processing of terrestrial measurements by "Systeem Detailmeting '76" (DUT)
AS 12.23.09/12.23.13
Key words: Cadastre, topographic base map

In the Cadastral Service terrestrial measurements are taken for fixing of cadastral boundaries of right and for collecting data for topographical base maps. A description is given of the system that has been developed.

The handling of the system can take place centrally or locally. The system runs both on the central computer of the Cadastre (VAX 11/780) and on the table computers at the provincial branch offices (Tektronic 4051/52).

(Author's summ.)

Pouls, National Correspondent of the Netherlands IGD 84/II 69

528.023.3:528.531

Reznickijj, F. E.
Vychislenie prevyshenijj i gorizontal'nykh prolozhenijj linijj, izmerennykh ehlektronnymi takheometrami
Geod. i fotogramm. gorn. dele. — Sverdlovsk (1982) 9. — p. 33—39 : 4 lit.
Computation of height differences and ground distances of lines measured by electronic tacheometers (RUS)
AS 12.23.09/73.07
Key words: Tacheometer, tacheometry

Dobrokhotov, National Correspondent, USSR IGD 84/II 70

528.42:528.53:681.3

Urban, A.
Kaartvervaardiging met het Wild Geomat-systeem bij Ingenieursbureau Oranjewoud BV
Geodesia Nederl. geod. T. — Apeldoorn 24 (1982) 7/8. — p. 251—255 : 7 fig. 3 tab. 2 lit.
The Wild Geomat interactive graphics and surveying system (DUT)
AS 12.23.09
Key words: Geomat, surveying system

The hardware components plus the software package (including the coding) of the Geomat system as it is being used by *Ingenieursbureau Oranjewoud BV* are described. *Oranjewoud* added testing procedure and utility programs to the software package. To illustrate the possibilities of the system, three different kinds of projects are presented.

(Author's summ.)

Pouls, National Correspondent of the Netherlands IGD 84/II 71

AIDOS Systematics **12.23.13** **Group II Surveying and topography**

Topographic and engineering-geodetic maps

Content of large-scale topographic and engineering-technical maps and plans and their assignment to definite purposes
Land maps
Symbols and methods of representation
Automation of the production and revision of large-scale topographic, engineering-technical maps and plans as well as of land maps (on the basis of geodetic methods of information acquisition)
Digital terrain models

Gruppe II Angewandte Geodäsie und Topographie

Topographische und ingenieurgeodätische Karten

Inhalt und Zweckbestimmung großmaßstäbiger topographischer und ingenieurtechnischer Karten und Pläne
Liegenschaftskarten
Signaturen und Darstellungsverfahren
Automatisierung der Herstellung und Laufendhaltung großmaßstäbiger topographischer, ingenieurtechnischer Karten und Pläne sowie von Liegenschaftskarten (auf der Grundlage geodätischer Verfahren der Informationsgewinnung)
Digitale Geländemodelle

528.22
528.42:681.3

Bashich, T. et al.
Metodologija tvorbe „temeljnih" digitalnih modela reljefa (TDMR) za potrebe fizikalne geodezije
Geodet. List. – Zagreb 36 (1982) 4–6. – p. 111–120 : 2 fig. 16 lit.
Methodology of establishing fundamental digital relief models for purposes of physical geodesy (GER)
AS 12.23.13/12.15.09
Key words: Digital terrain model, physical geodesy, Yugoslavia

Determination of initial criteria and elaboration of the structure of these relief models. Problems of their adaptation to different fields of application.

Kombinat Geod./Kart. IWT, Leipzig IGD 84/II 72

528.425.2(084.3-11):528.531

Boot, L. P.
Tachymetrie bij Fugro BV
Geodesia Nederl. geod. T. – Apeldoorn 24 (1982) 7/8. – p. 267–271 : 13 fig.
Tacheometry by Fugro BV (DUT)
AS 12.23.13
Key words: Large scale map, tacheometer, tacheometry

Since 1978 "*Ingenieurbureau Fugro BV*" uses the Intergraph interactive graphical system for processing tacheometric measurements for large scale map. Procedures from measurement and coding-system by self-registering tacheometers to resulting digital terrain model are described. The data base management and retrieval system used is explained.

(Author's summ.)

Pouls, National Correspondent of the Netherlands IGD 84/II 73

528.425
528.936.001.04:681.31

Divenyi, P.
Kiserlet terepi valtozasok ertekelesere
Geod. es Kartogr. – Budapest 35 (1983) 2. – p. 97–103 : 3 fig. 3 tab. 22 lit.
Experiment to evaluate changes in the terrain (HUN)
AS 12.23.13/12.33.25
Key words: Control of terrain changes, map improvement, map revision

The experiment described in the study has established quantitative and qualitative terrain changes in lowland areas of Hungary. The method applied was the determination of changes on the basis of a direct way, using a model structure. The author in the study compares the changing parameter evaluated cartometrically from the map (direct method) with other parameters taken from statistical pub-

lications (indirect method), and the two methods — according to the research — proved to be adequate.
This fact may have importance in the theoretical establishment of differentiated map revision system.
Veroe-Hetenyi, National Correspondent, Hungary IGD 84/II 74

528.42(084.3-11/.12)(430.1)

. . .
Funktion und Gestaltung der Deutschen Grundkarte 1 : 5000
Vermess. Photogramm. Kulturtechnik. — Zuerich 81 (1983) 7. — p. 252—254 : 1 lit.
Function and design of the German Basic Map 1 : 5,000 (GER)
AS 12.23.13
Key words: Basis map, FRG, large-scale map
Techn. Univ. Dresden IGD 84/II 75

528.425(492):528.711.1(203)
Linden, J. A. van der
Kartografie vanuit de lucht; de luchtfoto als informatiebron voor de topografische kaarten
Geodesia Nederl. geod. T. — Apeldoorn 24 (1982) 9. — p. 300—312 : 8 fig. 16 lit.
Cartography from the air; aerial photographs — a source for topographical mapping of the Netherlands (DUT)
AS 12.23.13/12.33.37
Key words: Aerial photograph, topographical map, the Netherlands
Since 1932 the official Dutch topographical maps, produced by the Topographic Service (TDN), have been based in photogrammetry. Even earlier than this aerial photography was used but only to expedite terrain reconnaissance.
The methods applied by the TDN differ from most other government map producing agencies. In particular there has been a consistent use of single plate photogrammetry applying direct scribing of the photographic imagery itself.
The keywords are: aerial triangulation, block adjustment, orthophoto production, direct scribing. Because of the increasing need for digital topographic data digital mapping is to be introduced gradually.
Pouls, National Correspondent of the Netherlands IGD 84/II 76

528.425:681.3
Markus, B.
Digitalis domborzatmodellezes
Geod. es Kartogr. — Budapest 35 (1983) 1. — p. 21—25 : 33 lit.
Digital terrain modelling (HUN)
AS 12.23.13
Key words: Digital terrain model, geodetic data processing, topography
Meaning and sense of digital terrain models are discussed, their applications surveyed and methods in establishing them reviewed. Trends of future developments are outlined.
(Author's summ.)
Veroe-Hetenyi, National Correspondent, Hungary IGD 84/II 77

528.425(086.5):681.3.04
Sanda, Vaclav / Novak, Pavel
Gestaltung und Ausnutzung des digitalen Gelaendemodells
Sbornik vyzkumn. praci VUGTK, R. 3. — Praha (1982) 14. — p. 149—169 : 6 fig. 1 tab. 8 lit.
Processing and application of digital terrain model (GER)
AS 12.23.13/12.23.21
Key words: Digital terrain model
Techn. Univ. Dresden IGD 84/II 78

528.425:681.3.04

Zinger, V. E.
Chastnaja klassifikacija geodezicheskikh modelejj i ikh sravnitel'nyjj analiz
Geod., Kartogr. i Aehrofotos"jomka. — L'vov (1982) 36. — p. 50—56 : 8 lit.
A partial classification of geodetic models and their comparative analysis (RUS)
AS 12.23.13/12.15.37
Key words: Digital terrain model

A classification of geodetic models based on particular features of representation of initial field data is presented. A comparative analysis of principal properties of mathematical, image-signatural and digital terrain models necessary for a successful choice of a particular class for corresponding studies has been carried out.
Dobrokhotov, National Correspondent, USSR IGD 84/II 79

AIDOS Systematics **12.23.21** **Group II Surveying and topography**

Application of geodesy in construction engineering and in traffic engineering

Engineering surveying operations in the designing, construction and using of industrial, civil-engineering, residential and public buildings
Reconnaissance, setting-out and route- and location surveys
Deformation measurements, subsidence and settlement measurements
Methods and means of engineering surveying

Gruppe II Angewandte Geodäsie und Topographie

Anwendung der Geodäsie im Bau- und Verkehrswesen

Ingenieurgeodätische Arbeiten bei der Projektierung, Errichtung und Nutzung von Industrie-, Ingenieur-, Wohn- und Gesellschaftsbauten
Erkundungen, Absteckungs- und Trassierungsarbeiten
Deformationsmessungen, Setzungs- und Senkungsmessungen
Methoden und Mittel der Ingenieurgeodäsie

528.48

Allan, A. L.
Present developments in high accuracy surveying in engineering and industrial sites (ENG)
Survey Rev. — Tolworth 27 (1983) 209. — p. 125—136 : 17 fig. 8 lit.
AS 12.23.21
Key words: Engineering surveying

In this paper, the author writes about specialized equipment and techniques which are used to solve measurement and alignment problems arising in typical engineering and industrial fabrication and construction. He also shows how conventional surveying equipment is often unsuited to perform these tasks with sufficient speed and accuracy. A brief speculation on future trends is included.
Cotter, National Correspondent, of Great Britain IGD 84/II 80

528.414
528.02:528.48

Atakishiev, A. A.
Primenenie geodezicheskikh metodov pri analize svobodnykh kolebanijj slozhnykh sistem
Ukr NIINTI Nr. 24Ui-DEP83. — 13 p. : 5 lit.
The use of geodetic methods in the analysis of free oscillations of compound systems (RUS)
AS 12.23.21/12.15.33
Key words: Engineering surveying
Dobrokhotov, National Correspondent, USSR IGD 84/II 81

528.486:625.11

. . .

Beitraege zur grossraeumigen Neutrassierung
Schr.-R. Wiss. Studiengang Vermess.-Wesen Hochsch. Bundeswehr Muenchen. — Muenchen (1982) 6. — 268 p.
(hrsg. v. Caspary, W. u. Welsch, W.)
Contributions to large-space relocation (GER)
AS 12.23.21
Key words: Digital terrain model, fast train systems, railway location

After an introduction on electromagnetic fast train systems, problems of preliminary planning, conception, technology, and railway location are discussed in 11 contributions. Studies of environmental effects, computer-assisted lay-out and optimization of routes, application of programme systems, use of graphic representation techniques, and documentation of results.

Techn. Univ. Dresden IGD 84/II 82

528.421:624

Cyrklaff, Gerhard
Zur Bestimmung und Anwendung koordinatenmaessig bekannter Gebaeudedecken
Vermessungstechnik. — Berlin 32 (1984) 1. — p. 11—12 : 2 fig. 7 tab. 2 lit.
The determination and application of co-ordinates of corners of structures with (GER)
AS 12.23.21
Key words: Position measurement

On building sites approximately 30 per cent of geodetic ground control points are destroyed as a consequence of building work. In the reconstruction of the control point net it is economical to restrict the remarking of ground control points to a minimum number, and to determine the co-ordinates of corners of structures which were built in the meantime. By means of the polar method with an electro-optical direction-distance measuring device and an auxiliary apparatus with single-prism reflector a standard deviation in position s_L of 5 to 10 mm is obtained. (Author's summ.)

Techn. Univ. Dresden IGD 84/II 83

528.48:624.9

Dvoreckijj, N. P.
Vertikal'noe proektirovanie pri geodezicheskom obespechenii stroitel'stva reaktornogo otdelenija AEhS
Ehnerg. str.-vo. — (1982) 9. — p. 53—55
Vertical planning in establishing geodetic control for construction of the reactor section of an atomic power-station (RUS)
AS 12.23.21
Key words: Atomic power-station, engineering surveying

Dobrokhotov, National Correspondent, USSR IGD 84/II 84

528.48:624/625

Fleischer, Ekkard
Beitrag zur Ermittlung der Massgenauigkeit im Stahlbetonskelettbau
Diss. TU Braunschweig
Geod. Schr.-R. Techn. Univ. Braunschweig. — Braunschweig (1979) 2. — 120 p. : div. fig. div. tab. div. lit.
(s. a. in Allg. Vermess.-Nachr. — Karlsruhe 87 (1980) 10. — p. 422)
On the determination of measuring accuracy in reinforced concrete skeleton constructions — Diss. TU Braunschweig (GER)
AS 12.23.21
Key words: Engineering surveying, skeleton construction

Surveying procedures to determine accuracy and to set out buildings continue to be devised and tested with regard to their suitability for skeleton constructions.

Determination of accuracy in production and erection is based on the author's own extensive accuracy tests on prefabricated parts and newly erected skeleton constructions as well as on studies of other agencies already published. The degree of accuracy necessary in setting out was determined according to the author's own theoretical studies.

Overall the studies showed that in particular for rod-shaped components and for horizontal distance accuracy and vertical alignment accuracy a revision of standard specifications should be undertaken to adjust these to the degree of accuracy obtainable at present.

Schoen, National Correspondent, Federal Republic of Germany IGD 84/II 85

528.486

Gozhijj, A. V.
Obshhijj princip ocenki tochnosti detal'nojj razbivki krugovojj krivojj razlichnymi sposobami
Geod., Kartogr. i Aehrofotos"jomka. — L'vov (1982) 36. — p. 24—33 : 4 lit.
A general principle of estimation of the accuracy of detailed setting-out of a circular curve by different methods (RUS)
AS 12.23.21
Key words: Setting-out curves

Dobrokhotov, National Correspondent, USSR IGD 84/II 86

528.482:691.3

Herda, Miroslav
Kontrollmessungen geometrischer Parameter raeumlicher Bauelemente
Sbornik vyzkumn. praci VUGTK, R 3. — Praha (1982) 14. — p. 137—147 : 6 fig. 7 lit.
Check measurement of geometric parameters of room-cell elements (GER)
AS 12.23.21
Key words: Building industry, Engineering geodesy

The prefabrication remains always the main factor of further rationalisation in building industry. The next stage of prefabrication in Czechoslovakia presents the assembly of structures composed of room-cell elements with dimension $2.4 \times 3.2 \times 9.8$ m. This prefabrication compared with frame constructions or panels much greater precision in setting-out as well as for assembly and inspection. The tolerances for the mentioned room-cell elements were determined to demands be 12 mm in the length, 5 mm in the width and 6 mm in the height. In accordance with the CMEA Standard for check measurement of geometric parameters in building industry the standard deviations in value ± 1.0 mm in the length, and ± 0.5 mm in the width and in the height result of this dimensional tolerances. A check measuring method with three theodolites standpoints in the same hight level was developed for the inspection of this elements.

(Author's summ., abbr.)

Techn. Univ. Dresden IGD 84/II 87

528.32:528.512.089.6

Horvath, K. / L'Aune, O.
Oesszehasonlito alapvonalak szabatos meghatarozasa a paksi atomeroemue epitesenel
Geod. es Kartogr. — Budapest 35 (1983) 1. — p. 11—17 : 2 tab. 18 lit.
Precise determination of comparison base lines at the Paks Nuclear Power Station (HUN)
AS 12.23.21/73.07
Key words: Base line, calibration of measuring tapes, statistical calibration

The exact length-determination of five comparator bases with different lengths and the same starting point established in the site of a building at an atomic power station is presented.
The accurate, precise length determination of the normal meter pair used in the measurements were performed by the so-called "statistical calibration" method, and making use of the technical aid of the National Office of Measures, by laser interferometer and universal length-measuring engine. The method of statistical calibration is also discussed in detail. The core of the method is the decrease of the constant in the equation of the normal meter and the efficient increase in the actual highest accuracy at the same time.
(Author's summ., abbr.)
Veroe-Hetenyi, National Correspondent, Hungary IGD 84/II 88

528.486.063:514.122.2
Kahler, Dieter
Ein Beitrag zur Berechnung der kartesischen Koordinaten von Klotoidenkleinpunkten
Z. Vermess.-Wesen. — Stuttgart 105 (1980) 5. — p. 232—240 : 5 fig. 5 lit.
A contribution to calculation of the Cartesian co-ordinates of clothoid detail points (GER)
AS 12.23.21
Key words: Cartesian co-ordinates, clothoid
A method to compute Cartesian co-ordinates of clothoid detail points by solving the Fresnel integrals using numerical integration is discussed. In addition to a comparison of different formulas an estimation of residual errors is given.
Schoen, National Correspondent, Federal Republic of Germany IGD 84/II 89

528.024.4:528.482
Kalugin, Ju. V.
Ispol'zovanie trigonometricheskogo nivelirovanija dlja nabljudenijj za osadkami sooruzhenijj
Geod. kontrol' v str-ve. — Kujjbyshev (1982). — p. 65—67
The use of trigonometric levelling for monitoring of construction settings (RUS)
AS 12.23.21/12.15.33
Key words: Trigonometric levelling
Dobrokhotov, National Correspondent, USSR IGD 84/II 90

528.482:621.311.25
Karsay, F. / Specuoar, A.
A Paksi Atomeroemue epitesenel vegzett geodeziai munkak eddigi tapasztalatai
Geod. es Kartogr. — Budapest 35 (1983) 2. — p. 89—92 : 4 fig.
Some experiences with surveying of the construction of the atomic energy plant Paks (HUN)
AS 12.23.21
Key words: Atomic energy plant, engineering geodesy, surface movement observation
A review is given on different kinds of surveying in connection with the construction of an atomic energy plant and its sections.
Veroe-Hetenyi, National Correspondent, Hungary IGD 84/II 91

528.482:625.72:681.3
Karsay, L.
Szamitogep alkalmazasa az uttervezesben
Geod. es Kartogr. — Budapest 35 (1983) 2. — p. 93—97
AS 12.23.21/63.23
Key words: Computer, earth works, road design

The UVATERV (Consulting and Engineering for Communication and Transportation) has participated in computer-aided design for 20 years. The engineers have reached the most considerable progress in the field of roads and motorways planning. They have a complete programme-system in this field. The article reviews the solutions of planning problems in seven topics as follows:

1. Horizontal alignment
2. Terrain data
3. Vertical alignment
4. Control of horizontal and vertical alignments
5. Cross-sections
6. Calculation and distribution of earthworks
7. Calculation of setting out data.

Veroe-Hetenyi, National Correspondent, Hungary IGD 84/II 92

528.48

Kirchner, Wolfram / Cremer, Konrad
Grossbauvorhaben und ihre (Ingenieur-)Vermessung
Allg. Vermess.-Nachr. — Karlsruhe 90 (1983) 8—9. — p. 335—342 : 7 fig. 2 lit.
Large building projects and their (engineering) surveying (GER)
AS 12.23.21/12.23.25/12.23.29
Key words: Engineering surveying

Techn. Univ. Dresden IGD 84/II 93

528.486:72

Kostina, G. D.
Rezul'taty teoreticheskikh i ehksperimental'nykh issledovanijj tochnosti geodezicheskikh razbivochnykh rabot i montazha ehlementov polnosbornykh zdanijj
VINITI Nr. 1527-83 Dep. — 53 p. : 9 lit.
The results of theoretical and experimental studies of the accuracy of geodetic setting-out operations and assembly of elements of fully-assembled constructions (RUS)
AS 12.23.21
Key words: Engineering geodesy, setting-out

Dobrokhotov, National Correspondent, USSR IGD 84/II 94

528.48:66.041.57

Krystowczyk, B.
Forgokemencek tengelyenek helyzetmeghatarozasa es beszabalyozasa
Geod. es Kartogr. — Budapest 35 (1983) 3. — p. 178—184 : 8 fig. 2 tab. 1 lit.
Positioning and regulation of the axis of rotary furnaces (HUN)
AS 12.23.21
Key words: Axis of rotary furnaces, furnace, piecometer, positioning of the axis

Determining the position of the axis of rotary furnaces in full operation has not yet had precedents. By presenting the results of actual experiments the author proves the truth of his statement, that the spatial position of the axis of a rotary furnace under operating circumstances differs from when it is at a standstill. The reasons are analysed in details. The special instrument designed just for this purpose and the measurement technique are presented. The technology yielding excellent results is adopted in factories all over Poland.

Veroe-Hetenyi, National Correspondent, Hungary IGD 84/II 95

528.48:624.9

Kuz'min, G. I.
Kontrol'nye izmerenija kachestva izgotovlenija bol'sheproljotnykh form geodezicheskimi metodami
Geod. kontrol' v str-ve. — Kujjbyshev (1982). — p. 76—79
Geodetic measurements for quality tests of wide-span trusses (RUS)
AS 12.23.21
Key words: Engineering surveying, supporting structure

Dobrokhotov, National Correspondent, USSR IGD 84/II 96

528.48

Matveev, S. I.
Voprosy inzhenernojj geodezii (po materialam seminara)
Geod. i Kartogr. — Moskva (1982) 11. — p. 23—25
The problems of engineering geodesy (seminar proceedings) (RUS)
AS 12.23.21/12.23.29
Key words: Engineering surveying

Dobrokhotov, National Correspondent, USSR IGD 84/II 97

528.48:624.9

Roshhupkin, V. D. / Shumkov, A. K.
Vyverka sooruzhenijj bashennogo tipa bol'shikh razmerov
Inzh.-geod. raboty pri str-ve, montazhe i ehkspluat. inzh. sooruzh. — (1982). — p. 73—79
Geodetic checking of large tower-type structures (RUS)
AS 12.23.21
Key words: Engineering surveying, tower-building

Dobrokhotov, National Correspondent, USSR IGD 84/II 98

528.541:528.48

Ruskov, A. M.
Rasshirenie vozmozhnostejj primenenija nivelira NT v inzhenerno-stroitel'-nykh izyskanijakh
Geod. kontrol' v str-ve. — Kujjbyshev (1982). — p. 90—91
Expansion of application possibilities of the NT level in engineering reconnaissance surveys (RUS)
AS 12.23.21/73.07
Key words: Engineering geodesy, level

Dobrokhotov, National Correspondent, USSR IGD 84/II 99

528.548:528.482

Rylance, K. A.
Karyplummet, a vertical reference line (ENG)
Survey Rev. — Tolworth 26 (1982) 204. — p. 295—298 : 4 lit.
AS 12.23.21/73.07
Key words: Plummet, tower movement

The equipment here described was devised to monitor the movements of a bell tower whilst the bells were being moved. Subsequently the equipment has been used for many purposes where the definition of a vertical reference line has been required. Many of the problems experienced with optical devices, plumb bobs and lasers have been solved.
528.541:528.48

Cotter, National Correspondent, of Great Britain IGD 84/II 100

528.48

Shiba, M. A. / Ozhigov, S. A.
Geodezicheskie raboty pri vozvedenii telebashni
Geodez. i Kartogr. — Moskva (1982) 9. — p. 21—24
Geodetic survey when constructing a television tower (RUS)
AS 12.23.21
Key words: Engineering surveying
Dobrokhotov, National Correspondent, USSR IGD 84/II 101

528.486:625.143

Stephan, Margit
Abstecken von Gleisbauanlagen mit einer neuen Pfeilhoehenmesseinrichtung
Vermessungstechnik. — Berlin 32 (1984) 1. — p. 9—10 : 2 fig. 1 lit.
Location of track systems with a new device for measuring the sag (GER)
AS 12.23.21
Key words: Setting out railway trackage

Description of a device for measuring the sag by which the efficiency is increased considerably and the working conditions are improved. This measuring device consists of a track tripod, theodolite, measuring rod, and sighting rod. The demanded accuracy of a measured sag $m_h = \pm 1$ mm is kept. The measuring point is the bottom edge of the inner side of the rail-head. A measuring technology was elaborated for performance of this measurement.
(Author's summ.)
Techn. Univ. Dresden IGD 84/II 102

528.486:625.11
528.531

Triceckijj, N. N.
Primenenie takheometrov dlja obmera zheleznodorozhnykh krivykh
Geodez. i Kartogr. — Moskva (1982) 9. — p. 30—31
The use of tacheometers for measuring railway curves (RUS)
AS 12.23.21/73.07
Key words: Measurement of curves, railway surveying, tacheometer
Dobrokhotov, National Correspondent, USSR IGD 84/II 103

528.48:528.526.6:625.72

Zhilin, S. N.
Metodika obsledovanija trass avtomobil'nykh dorog s ispol'zovaniem giroskopicheskojj apparatury
Sb. nauch. tr. Gos. dop. proekt.-izyskat. i NII. — (1981) 33. — p. 77—80
The methods of reconnaissance of express highways routes by using gyroinstruments (RUS)
AS 12.23.21/73.07
Key words: Express highway, gyroscopic instrument
Dobrokhotov, National Correspondent, USSR IGD 84/II 104

AIDOS Systematics **12.23.29 Group II Surveying and topography**

Application of geodesy in agriculture, forestry and water resources management and for environmental protection

Geodetic work for land management and the preparation of large-scale ground maps, forestry maps and water-resources-management maps
Ground cadastre
Forest taxation

Gruppe II Angewandte Geodäsie und Topographie

Anwendung der Geodäsie in der Land-, Forst- und Wasserwirtschaft sowie beim Umweltschutz

Geodätische Arbeiten für die Landeinrichtung und Herstellung von großmaßstäbigen Boden-, Forst- und wasserwirtschaftlichen Karten
Bodenkataster
Forsttaxation

528.482:627.8

Cjun'ko, V. I.
Prognoz ehkstremal'nojj osadki gidrosooruzhenija, vyzvannojj sezonnymi izmenenijami urovnja vody v vodokhranilishhe
Geod. i fotogramm. — Rostov n/D. (1981). — p. 66—71 : 2 lit.
Prediction of maximum settlings of hydroelectric constructions due to seasonal water-level variations in the reservoir (RUS)
AS 12.23.29
Key words: Engineering geodesy, hydroelectric power station

Dobrokhotov, National Correspondent, USSR — IGD 84/II 105

528.024.4:551.49

Krueger, Helmut / Siedler, Ulrich
Anwendungsmoeglichkeiten des trigonometrischen Nivellements in der hydrogeologischen Erkundung
Vermessungstechnik. — Berlin 32 (1984) 1. — p. 19—22 : 4 tab. 9 lit.
Applications of trigonometrical levelling in hydrogeological research (GER)
AS 12.23.29
Key words: Geology, trigonometrical levelling

Establishment of a measuring baseline of length 1,050 m subdivided into 8 partial sections. Measurement of differences in height in all combinations and analysis of result. Description of the advantages of trigonometrical levelling in hilly terrain. Trigonometrical levelling should be used in position determinations with electro-optical tachymeters. The mean error per kilometre of different measuring methods amounts to approx. ± 6 to 8 mm. The most favourable sighting range is 200 to 600 m.

(Author's summ.)

Techn. Univ. Dresden — IGD 84/II 106

International Union of Geodesy and Geophysics
International Association of Geodesy

bibliographia geodaetica

AIDOS Systematics **12.29** **GROUP III PHOTOGRAMMETRY AND REMOTE SENSING**

GRUPPE III PHOTOGRAMMETRIE UND FERNERKUNDUNG

AIDOS Systematics **12.29.05** **Group III Photogrammetry and remote sensing**

Fundamentals of photogrammetry and remote sensing

Mathematical basis of photogrammetry
Affine and projective descriptive geometry for photogrammetric processes
Optical and photogrammetric imaging
Stereoscopic vision and measurement
Types and properties of photograms

Gruppe III Photogrammetrie und Fernerkundung

Grundlagen der Photogrammetrie und Fernerkundung

Mathematische Grundlagen der Photogrammetrie
Affine und projektive darstellende Geometrie für photogrammetrische Prozesse
Optische und photogrammetrische Abbildungen
Stereoskopisches Sehen und Messen
Arten und Eigenschaften der Meßbilder

528.711.7:377/378
681.32:528.8

Eyton, J. Ronald
A hybrid image classification instructional package (ENG)
Photogramm. Engin. & Remote Sens. — Falls Church, Virg. 49 (1983) 8. — p. 1175—1181 : 7 fig. 6 lit.
AS 12.29.05/63.11
Key words: Digital image processing, education, image classification

An introductory untergraduate course in digital remote sensing has been successfully conducted using a combination of Landsat image processing programs and mainframe statistical and mapping library routines. This hybrid approach allows for considerable flexibility and makes maximum use of the software and hardware commonly supported by an academic computer center. Students work through the entire classification process starting with Landsat image subsets extracted from a Computer Compatible Tape (CCT) and finishing with a colored hard copy map of the final classification. Emphasis is placed on students understanding the various classification procedures at the expense of image processing efficiency. A number of different computer graphic algorithms are used to facilitate the students' comprehension of both the spatial and statistical content of the image data.
(Author's summ., abbr.)

Techn. Univ. Dresden IGD 84/III 53

528.711.112:535.316/.317

Hildebrand, Klaus
New generation lens systems for the Wild Aviophot camera system (ENG)
Photogramm. Engin. & Remote Sens. — Falls Church, Virg. 49 (1983) 8. — p. 1201—1209 : 13 fig. 4 tab. 5 lit.
AS 12.29.05/12.29.13/73.11
Key words: Aerial-camera lens, AWAR, image quality, MTF

This contribution reports on some of the criteria which have to be met in the further development of aerial-camera lens systems. The quality criteria that were constantly observed in the development work are described in detail, among which integrals on the MTF and their area-weighted average proved particularly useful. Details are given of the metric characteristics and of the image quality of the new lens systems in the visible and near infrared range, and these are compared with the comparable performance data of the older lens types.

It is demonstrated that the lens systems of this new generation have a substantially greater information-transmission capacity, a more homogeneous image quality, and lower residual distortion.

(Author's summ.)

Techn. Univ. Dresden IGD 84/III 54

528.711.7:681.32
377/378:528.8

Kiefer, Ralph W. / Gunther, Fred J.
Digital image processing using the Apple II microcomputer (ENG)
Photogramm. Engin. & Remote Sens. — Falls Church, Virg. 49 (1983) 8. — p. 1175—1167—1174
AS 12.29.05/63.11
Key words: Digital image processing, education, microcomputer

The inexpensive Apple II microcomputer can be used for digital image processing in operational and educational environments. The Apple II is portable enough to be taken directly into the classroom for lecture demonstrations and laboratory exercises. It is easily taken into the field for on-site data analysis or local presentations. Processing results can be displayed on a color monitor/TV screen in either low-resolution or high-resolution graphics, or can be printed on a line printer. At least five digital image processing systems (software packages) are in use in the United States: Apple II Digital Image Processing System (A/DIPS), Apple Image Processing Educator (AIPE), Oklahoma Landsat Training Program System, Apple Personal Image Processing System (APPLEPIPS), and Mini-LARSYS.

(Author's summ.)

Techn. Univ. Dresden IGD 84/III 55

528.7(492)(091)

Weele, A. J. van der
Enkele hoofdpunten van de fotogrammetrie in Nederland in de periode 1957—1982
Geodesia Nederl. geod. T. — Apeldoorn 24 (1982) 9. — p. 286—294
Some highlights from the history of photogrammetry in the Netherlands (DUT)
AS 12.29.05/12.01.09
Key words: History, the Netherlands, photogrammetry

The author describes some of the main developments in the Netherlands during the second periods of the existence of the Netherlands Society for Photogrammetry. Attention is mainly paid to the technical development during this period.

In connection with the introduction of automation in photogrammetry some words are devoted to the social aspects of this development.

(Author's summ.)

Pouls, National Correspondent of the Netherlands IGD 84/III 56

528.8:377/378
681.32:528.711.7

Williams, T. H. Lee / Gunn, Christopher / Siebert, Jeffrey
Instructional use of a mainframe interactive image analysis system (ENG)
Photogramm. Engin. & Remote Sens. — Falls Church, Virginia 49 (1983) 8. — p. 1159—1165 : 4 fig. 3 lit.
AS 12.29.05/63.11
Key words: Digital image processing, education, image analysis

The increasing demand for training in digital image analysis techniques calls for the widespread development of suitable training courses in academic institutions. Implementation of image analysis program packages on college and university mainframe computers can aid this development in a cost-effective manner. The Department of Geography-Meterology and the Kansas Applied Remote Sensing (KARS) Program have developed an instructional interactive image analysis program package that runs on the University of Kansas central computer. A brief description of the program package is given, followed by a discussion of the advantages and limitations of image analysis instruction in a time-sharing mainframe environment.

(Author's summ.)

Techn. Univ. Dresden IGD 84/III 57

061.3(100+44)"1982"ISPRS/Comm. VII:528.7

Wilmet, J.
Le symposium international de la Commission VII de la Societe Internationale de Photogrammetrie et de Teledetection, Toulouse, 13—17 Septembre 1982
Bull. trimestr. Soc. belge Photogramm. et de Teledetec. — Bruxelles (1983) 149—150. — p. 9—12
The international symposium of Commission VII of the International Society of Photogrammetry and Remote Sensing, Toulouse, Sept. 13 to 17, 1982 (FRE)
AS 12.29.05/12.01.17
Key words: International Society of Photogrammetry, symposium 1982

From a great variety of contributions the author derives definite trends of development and gives some suggestions for the future congress in Rio de Janeiro, in 1984.

Techn. Univ. Dresden IGD 84/III 58

AIDOS Systematics **12.29.09** **Group III Photogrammetry and remote sensing**

Terrestrial photogrammetry

Methods of terrestrial survey
Use of instruments in terrestrial photogrammetry

Gruppe III Photogrammetrie und Fernerkundung

Terrestrisch-photogrammetrische Aufnahme

Verfahren der terrestrisch-photogrammetrischen Aufnahme
Nutzung von Geräten der terrestrisch-photogrammetrischen Aufnahme

528.721.1.088(21)

Elorabi, Abdelmoniem

Empirische Genauigkeitsuntersuchung im terrestrisch-photogrammetrischen Einzelmodell — Diss.

Karlsruhe, Univ., Fak. Bauing.- und Vermessungswesen, Diss., 1983. — 136 p. : 27 fig. 42 tab. 49 lit.

Empirical accuracy investigation in the terrestrial-photogrammetric single model — diss. (GER)

AS 12.29.09/12.29.25

Key words: Accuracy, correlation analysis, normal case, single model, terrestrial photogrammetry

A spatial test field forms the basis for the calibration of the measuring camera P 32 as well as for empirical accuracy investigations (120 models) with plates and film. Comparison of a two-step technique with the rigorous bundle solution. The use of plates and the application of the rigorous method proved to be favourable for a high accuracy of model co-ordinates. With consideration of empirical variances and covariances in the two-step solution (which were determined by the rigorous technique), considerable accuracy increases could be proved (70 per cent in the average) for model co-ordinates.

Techn. Univ. Dresden IGD 84/III 59

528.711.1(21):930.26

Jassemm, Khaleel I. / Scherz, James P.

Terrestrial photogrammetry as a tool for archaeoastronomical investigation (ENG)

Techn. Papers 1982 ACSM-ASP Fall Convention, Hollywood, Florida, September 19—23, 1982. Ed.: American Congr. Surv. & Mapp. / American. Soc. Photogramm. — Falls Church, Virginia 1982. — p. 221—230 : div. fig. div. lit.

AS 12.29.09/12.29.33

Key words: Archaeology, terrestrial photogrammetry

The purpose of this paper is to illustrate how terrestrial photography is being used to analyze factors at certain sites to ascertain whether or not they were of ancient sundial significance. Two 35 mm cameras were mounted on a specially designed tripod, which can provide the desired overlap for difference targets. To provide model control a special three-dimensional control grid was constructed which can be leveled and oriented to any direction. The resulting photography was analyzed by both semi-analytical and analytical methods and reasonable results were obtained. The dip and strike of rock faces were computed with a special mathematical model.

(Author's summ., abbr.)

Techn. Univ. Dresden IGD 84/III 60

AIDOS Systematics **12.29.13 Group III Photogrammetry and remote sensing**

Photogrammetric acquisition of information from the air

Using of equipment and techniques of air photography and of aerocosmic survey
Using of accessories
Survey aircraft and survey flight operations
Survey flight planning
Aerial photography

Gruppe III Photogrammetrie und Fernerkundung

Photographische Erfassung von Informationen aus der Luft

Nutzung von Ausrüstungen und Verfahren der Luftbild- und aerokosmischen Aufnahme
Nutzung von Hilfsgeräten
Bildflugzeuge und Bildflugarbeiten
Bildflugplanung
Aerophotographie

528.71.024(73):681.3

Antill, Paul A.
The national high-altitude photographic data base (ENG)
Techn. Papers 1982 ACSM-ASP Fall Convention, Hollywood, Florida, September 19—23, 1982. Ed.: American Congr. Surv. & Mapp. / American. Soc. Photogramm. — Falls Church, Virginia 1982. — p. 28—37 : 4 fig.
AS 12.29.13/12.29.25
Key words: High-altitude photography

The National High-Altitude Photography Program is a multiagency federal program to provide complete photographic coverage of the three million square miles their efforts and funds for a 6-year program that will save millions of dollars which comprise the coterminous United States. These agencies are combining through consolidating requirements and minimizing duplication. The consolidated requirements are met by the program's dual camera system which provides black-and-white photography for many cartographic uses and color-infrared photography for resource investigations involving soils, crops, timber, water quality, and farmland and wetland classification.

(Author's summ., abbr.)

Techn. Univ. Dresden IGD 84/III 61

528.721.122:624.131.7:622.323

Braverman, B. A.
Sovershenstrovanie tekhnologii analiticheskojj obrabotki aehrosnimkov pri izyskanijakh dlja neftepromyslovogo stroitel'stva
Neftepromys'l. str-vo. — Moskva (1981) 11. — p. 15—17
Improvement of technology of analytical processing of aerial photographs in reconaissance surveys for construction of oil-extracting enterprises (RUS)
AS 12.29.13/12.29.25/12.29.37
Key words: Investigation of foundation conditions, oil-conveyor equipment, photointerpretation, space photography

Dobrokhotov, National Correspondent, USSR IGD 84/III 62

528.711:771.531.37
528.74:551.3.053

Frazier, B. E. / Hooper, G. K.
Use of chromogenic film for aerial photography of erosion features (ENG)
Photogramm. Engin. & Remote Sens. — Falls Church, Virg. 49 (1983) 8. — p. 1211—1217 : 10 fig. 5 lit.
AS 12.29.13/12.29.25/65.33/51.47
Key words: Aerial photography, chromogenic film, low oblique photography, soil erosion

The potential of Ilford XP1 photographic film in 35-mm format for aerial photo assessment of erosion was investigated. A wide exposure latitude (ASA 50 to ASA 1600) makes this film attractive for aerial photography where lighting conditions are variable and rapid shutter speeds are required. The film was field tested for image quality using exposures equivalent to ASA 100 to ASA 1000. A D-log E curve was developed by photographing a gray card from ASA 40 to ASA 128,000. The D-log E curve from negatives has a long straight portion with 0.5 gamma and a toe that is barely perceptible beginning at ASA 1600. Acceptable prints were obtained throughout the range ASA 100 to ASA 1000. A bar chart photographed at ASA 1000 showed separable line pairs 25.4 mm in width, equivalent to 35 line pairs/mm on the negative. Examples show that the film is useful for documenting the effects of erosion after cultivation and for measuring rills or soil slips.

(Author's summ.)

Techn. Univ. Dresden IGD 84/III 63

528.711.11:681.3

Hecht, S.
CCD-Bilddatenverarbeitung mit Einchiprechner
Feingeraetetechnik. — Berlin 31 (1982) 9. — p. 414—416
CCD picture data processing using a monochip computer (GER)
AS 12.29.13/63.23
Key words: Data processing, measured value processing, photogram, programme (EDP)

By combining a high-performance monochip microcomputer in a CCD solid state camera, an independent processing of measured values of given video information is possible. Concerning circuit techniques low expenditure was aimed at, a flexible use of the camera was guaranteed and the capacity of the computer system was utilized.

Kombinat Geod./Kart. IWT, Leipzig IGD 84/III 64

528.711.4.029.67

Jansa, J.
Rektifizierung von Multispektral-Scanneraufnahmen
Geowiss. Mitt. TU Wien. — Wien (1983) 24. — 106 p.
Rectification of multispectral-scanner images (GER)
AS 12.29.13
Key words: Error detection, least squares interpolation, MSS images, rectification

It is a recurring demand of interpreters of scanner images to have these images geometrically corrected, and mounted as continuous mosaics. In accordance with the techniques of their creation, MSS images are distorted in complex ways, not found in conventional photography. Main types of distortion are discussed and illustrated by their influence on images of a regular grid.

A review of techniques to rectify MSS imagery densities, is an introduction to the main chapter. These are concerned with a practicable method to rectify MSS imagery using the nonparametric solution with least squares interpolation. The rectification can be performed digitally or optically (applying the Wild AVIOPLAN OR1).

(Author's summ., abbr.)

Erker, National Correspondent, Austria IGD 84/III 65

528.74:629.783:55

MacDonald, H. et al.
Shuttle imaging radar — geologic applications
Techn. Papers 1982 ACSM-ASP Fall Convention, Hollywood, Florida, September 19—23, 1982. Ed.: American Congr. Surv. & Mapp. / American Soc. Photogramm. — Falls Church, Virginia 1982. — p. 272—281 : 7 fig. 1 tab. div. lit.
AS 12.29.13/12.29.25/12.29.37
Key words: Landsat, radar imagery, Seasat, space shuttle

The Shuttle Imaging Radar-A (SIR-A), had as a prime objective to evaluate the capability of spaceborne imaging radars as a tool for geologic exploration. The results of the experiment will help to determine the value of using the combina-

tion of space radar and Landsat imagery for improved geologic analysis and mapping. Premiliminary analysis of the Shuttle radar imagery with Seasat and Landsat imagery from similar areas provides evidence that spaceborne radars can significantly complement Landsat interpretation, and vastly improve geologic reconnaissance mapping in those areas of the world that are relatively unmapped because of perpetual cloud cover.
(Author's summ., abbr.)
Techn. Univ. Dresden IGD 84/III 66

528.74:551.42

Souquiere, M.
L'exploitation hydrographique de photographies aeriennes prises a partir d'un aeronef localise par procede radioelectrique
Geometre. – Paris 126 (1983) 8–9. – p. 39–42 : 6 fig.
Hydrographic evaluation of aerial photographs except those taken by radiolocated aircraft (FRE)
AS 12.29.13/12.29.29/12.29.37
Key words: Hydrography, photogrammetric evaluation, photointerpretation

A photogrammetric method of boundary survey and topographic surveying of extended reef zones and atolls is illustrated with the example of New Caledonia.
Techn. Univ. Dresden IGD 84/III 67

528.711.11:681.3

Wurmus, H. / Hecht, S. / Richard, G.
Universeller Kopplungsmodul fuer Bildsensoren und Mikrorechnersysteme
Feingeraetetechnik. – Berlin 31 (1982) 9. – p. 411–414 : 8 lit.
Universal coupling module for picture sensors and microcomputer system (GER)
AS 12.29.13/63.23
Key words: Data processing, optical signal processing, photogram, photographic technique

Explanation of the initial technical conditions for designing the place of coupling between camera and microcomputer and description of the entire structure of the coupling module.
Kombinat Geod./Kart. IWT, Leipzig IGD 84/III 68

528.711:629.783:634

Zhirin, V. M.
Osnovy kosmicheskogo slezhenija za dinamikojj lesnogo fonda
Les. kh-vo. – (1982) 9. – p. 53–55 : 7 lit.
The principles of space monitoring of forest dynamics (RUS)
AS 12.29.13/12.29.33/87
Key words: Forestry, photointerpretation, space photography

Dobrokhotov, National Correspondent, USSR IGD 84/III 69

AIDOS Systematics **12.29.21 Group III Photogrammetry and remote sensing**

Photogrammetric determination of control points

Selection of control points for the photogrammetric and geodetic determination
Photogrammetric work for determining control points (as junction points for geodetic measurements)
Radial triangulation, aerial triangulation
Analytical aerotriangulation
Adjustment of strip and block triangulation

Gruppe III Photogrammetrie und Fernerkundung

Photogrammetrische Bestimmung von Paßpunkten

Auswahl von Paßpunkten für die photogrammetrische und geodätische Bestimmung
Photogrammetrische Arbeiten zur Bestimmung von Festpunkten (als Anschlußpunkte für geodätische Messungen)
Radialtriangulation, Aerotriangulation, analytische Aerotriangulation
Ausgleichung von Streifen- und Blocktriangulation

528.73(202)
Balashov, A. A. / Reshetov, E. A. / Savel'ev, B. I.
O rezul'tatakh fototriangulirovanija po kosmicheskim znimkam
Geod. i Kartogr. — Moskva (1982) 11. — p. 27—30
On the results of phototriangulation using space imagery (RUS)
AS 12.29.21/12.29.25
Key words: Phototriangulation
Dobrokhotov, National Correspondent, USSR IGD 84/III 70

528.738:681.3.001.1
Pietschner, Juergen / Schulz, Hans-Ulrich
Das Programmsystem TEBIT — Ueberblick und Grundkonzeption
Vermessungstechnik. — Berlin 32 (1984) 1. — p. 7—9 : 1 fig. 8 lit.
The programme system TEBIT — outline and fundamental conception (GER)
AS 12.29.21/12.29.25
Key words: Control points, terrestrial aerotriangulation
Terrestrial aerotriangulation has a great importance for the photogrammetric determination of control points or of new points. Modern techniques of analytical aerotriangulation permit the rational solution of problems of point determination appearing with more complex positions of exposure in close-range photogrammetry. The bundle method with the application of additional geodetic observations offers the best pre-requisites. The programme system TEBIT enables the evaluation of arbitrary systems of images (which consist of up to approx. 20 images) with a maximum of 500 new points. Description of fundamental structure and efficiency parameters.
(Author's summ.)
Techn. Univ. Dresden IGD 84/III 71

528.73:528.14:681.3
Schwarz, P. G.
FOTEF-Toetsing in de fotogrammetrie
Geodesia Nederl. geod. T. — Apeldoorn 24 (1982) 9. — p. 312—319 : 6 fig. 2 app. 7 lit.
FOTEF-testing in photogrammetry (DUT)
AS 12.29.21/12.29.25
Key words: Adjustment, photogrammetry
At the Survey Department of the Rijkswaterstaat (the public works depeartment) a program system (FOTEF) has been developed for the adjustment of independent model triangulations and for testing the data according to Baarda's date snooping method. A basic line of thought was the splitting up of the error detection into two parts: low cost pre-adjustment error detection first, followed by the search for smaller errors by testing statistically in the adjustment phase. The program consists of fast and low-cost pre-adjustment programs, programs for least squares adjustment and final statistical testing. FOTEF furthermore includes programs calculating precision and reliability data as required in quality studies.
(Author's summ.)
Pouls, National Correspondent of the Netherlands IGD 84/III 72

AIDOS Systematics **12.29.25** **Group III Photogrammetry and remote sensing**

Photogrammetric plotting

Fundamentals of the photogrammetric single-photograph plotting and double-image plotting
Graphical, optical-mechanical and analytical rectification, differential rectification, rectification with subjective and mechanical projection
Working with stereoplotting instruments (orientation)
Analytical plotting
Using of photogrammetric plotting instruments (comparators, analogue plotters, analytical plotting, stereoplotters, differential rectifiers, point transfer instruments)
Automation of the plotting processes
Digital recording of measurement results

Gruppe III Photogrammetrie und Fernerkundung

Photogrammetrische Auswertung

Grundlagen der photogrammetrischen Einzelbild- und Zweibildauswertung
Graphische, optisch-mechanische, analytische Entzerrung, Differentialentzerrung, Entzerrung mit subjektiver und mechanischer Projektion
Arbeit an Stereoauswertegeräten (Orientierung)
Analytische Auswertung
Nutzung von photogrammetrischen Auswertegeräten (Komparatoren, Analogauswertegeräte, analytische Auswertegeräte, Stereoauswertegeräte, Differentialentzerrungsgeräte, Punktübertragungsgeräte)
Automatisierung der Auswertungsprozesse
Digitale Erfassung der Meßergebnisse

528.721.126:528.236

Bormann, G. E. / Vozikis, E.
Map projection transformation with digitally controlled differential rectifiers (ENG)
Photogramm. Engin. & Remote Sens. — Falls Church, Virginia 49 (1983) 9. — p. 1317–1323 : 5 fig. 3 map. 3 lit.
AS 12.29.25/12.29.29
Key words: Orthophoto, projection transformation

The geometrically correct transformation of the entire map content from one cartographic projection to another is an old problem. It can be solved economically with digitally-controlled differential rectifiers such as the Wild AVIOPLAN OR1, using additional software modules. Examples illustrate the potential of this method and show also the great advantages of the color capability of the OR1 system.

(Author's summ.)

Techn. Univ. Dresden IGD 84/III 73

528.711.4:528.8.041.5(202)

Chizhevskijj, A. M.
Otdor informativnykh priznakov pri klassifikacii mnogozonal'nykh izobrazhenijj
Issl. Zemli Kosmosa. — Moskva (1982) 6. — p. 104–107
Selection of informative indications in classification of multispectral imagery (RUS)
AS 12.29.25/12.29.37
Key words: Multispectral image, photo-interpretation

An approach based on the Monte Carlo method allowing selection and ranging of indications for classification of multispectral imagery from computed mean probabilities of identification errors regardless of the distribution law of luminosity vectors of the classes given is discussed. An algorithm realizing the proposed method of selection of informative indications is given.

Dobrokhotov, National Correspondent, USSR IGD 84/III 74

528.74:72

Coster, A. de
Apport de la stereophotogrammetrie a l'etude de stabilite des edifices anciens
Bull. trimestr. Soc. belge Photogramm. et de Teledetec. — Bruxelles (1983) 149—150. — p. 13—21 : 6 fig. 10 lit.
Contribution of stereophotogrammetry to stability investigations of ancient structures (FRE)
AS 12.29.25
Key words: Architectural photogrammetry, stereophotogrammetry

By means of stereophotogrammetric images investigations on the deformation behaviour of a Gothic church are performed.

Techn. Univ. Dresden IGD 84/III 75

528.721.12
528.711.7:681.32

Dalia, O.
Digitalis kep geometriai transzformacioja
Geod. es Kartogr. — Budapest 35 (1983) 1. — p. 30—38 : 13 fig. 15 lit.
Geometric transformation of digital image (HUN)
AS 12.29.25/12.29.41
Key words: Control points, digital image transformation, LANDSAT images, remote sensing

Problem of geometrical distortion and correction of LANDSAT MSS and RBV images is presented. Basic internal and external distortions in case of both sensor type are illustrated. Type of transformation functions is polinomials. Transformation is determined by control points, using least square approximation method. Approximation technique is proposed to calculate corrected image, to reduce the computing time of computer programme. The elaborated method and FORTRAN programme can be used not only for LANDSAT image correction, and also in an important part of remote sensing applications program package, which is under development in the Institute of Geodesy and Cartography. Experimental results are presented and demonstration by uncorrected and corrected images is provided.

Veroe-Hetenyi, National Correspondent, Hungary IGD 84/III 76

528.721.2

Khajjdushki, Ivan T.
Afinni stereomodeli vahv fotogrametrijata
Geod., Kartogr., Zemeustrojjstvo. — Sofija (1983) 1. — p. 19—21 : 2 fig. 1 lit.
Affine stereomodels in photogrammetry (BUL)
AS 12.29.25
Key words: Stereomodel

The geometric relation between parallel projecting of two planes and perspective-affine transformation within the plane is examined. Some advantages and possibilities to determine the effective deformations of the affine stereomodels are shown both by terrestrial and by aerial analogous stereo-photogrammetry, in case of large scales.

Georgiev, National Correspondent, Bulgarian IGD 84/III 77

528.722.2:528.71(21)

Kis Papp, L.

Az UT 1 kepatalakito felhasznalasa foeldi fotogrammetriai felvetelek analog kiertekelesere

Geod. es Kartogr. — Budapest 35 (1983) 1. — p. 8—11 : 4 fig. 1 tab. 2 lit.

A transformation printer for terrestrial photogrammetry (HUN)

AS 12.29.25/12.29.37/73.11

Key words: Photogrammetric interpretation, terrestrial photogrammetry, transformation printer

After a review of the facilities for plotting of terrestrial photographs in existing analog plotters, the author describes the new Wild UT 1 transformation printer for terrestrial photographs. With its high optical and metric performance, this instrument provides the possibility of precise photographic enlargement with a fixed ratio of 2 of terrestrial pictures with short focal length, on film, or glass, in black-and-white or in colour. The transformed photographs are then compatible with the ranges of analog plotters.

Veroe-Hetenyi, National Correspondent, Hungary IGD 84/III 78

528.721.1:681.3

Klette, Reinhard

Parallelprozessoren fuer die digitale Bildverarbeitung

Teil I und II

Bild u. Ton. — Leipzig 36 (1983) 1. — p. 5—12, 26; 36 (1983) 2. — p. 37—43, 60 : 292 lit.

Parallel processors for digital image processing (GER)

AS 12.29.25

Key words: Analogue-digital conversion, data processing, optical signal processing, photograms, programming

The parallelization of data processing methods and the distribution of tasks within a computer system, represent indispensable principles for future digital image processing. The most essential basic principles of mastering computer structures working in a parallel way are developed so that a relatively extensive application of this computer technique in digital image processing will be practized soon in the years to come.

Kombinat Geod./Kart. IWT, Leipzig IGD 84/III 79

528.711.7(084.3):778.681.5

Likens, William / Maw, Keith

Updating Landsat-derived land-cover maps using change detection and masking techniques (ENG)

Techn. Papers 1982 ACSM-ASP Fall Convention, Hollywood, Florida, September 19—23, 1982. Ed.: American Congr. Surv. & Mapp. / American Soc. Photogramm. — Falls Church, Virginia 1982. — p. 256—271 : 3 tab. div. lit.

AS 12.29.25/65.31

Key words: Land-cover map, Landsat, masking technique

The use of Landsat data within a geographic data base is being investigated in a study known as the California Integrated Remote Sensing System's (CIRSS) San Bernardino Country Project.

An initial step in the process was to create a baseline land-cover classification, using a 1976 image. After performing geometric and radiometric adjustments to the 1976 image, four change-analysis methods were examined to compare the image with a 1979 scene. A mask of possible change areas between the base image and the 1979 update Landsat image was created, using the union of three of these techniques. The "change image" thus developed was used as a mask to limit analysis of the 1979 Landsat image.

(Author's summ., abbr.)

Techn. Univ. Dresden IGD 84/III 80

528.72

Lyskov, G. A. / Skosar', Ju. A.
Issledovanie vlijanija distorsii izobrazhenija na tochnost' opredelenija vysot v zavisimosti ot prodol'nogo perekrytija snimkov i sposobov postroenija modeli
Geod. i fotogrammetrija. — Rostov n/D. (1981). — p. 54—62
The study of the effect of photograph distortion on the accuracy of height determination depending on forward overlapping of photographs and the methods of model construction (RUS)
AS 12.29.25
Key words: Photogrammetric restitution, photograph distortion

Dobrokhotov, National Correspondent, USSR IGD 84/III 81

528.721.1:681.3

Pose, R. A.
Verfahren zum Auffinden und praezisen Vermessen von Isolinien auf Filmbildern
Bild u. Ton. — Leipzig 36 (1983) 2. — p. 56 : 15 lit.
Techniques of detection and precise surveying of isolines on film images (GER)
AS 12.29.25
Key words: Data processing, isoline, measuring image, optoelectronic device, scanning

Kombinat Geod./Kart. IWT, Leipzig IGD 84/III 82

528.7:513.012

Schmid, Hellmut H.
Analytische Photogrammetrie als ein Spezialfall der Ikonometrie
Vermess. Photogramm. Kulturtechnik. — Zuerich 81 (1983) 9. — p. 312—318 : 4 lit.
Analytical photogrammetry as a special case of iconometry (GER)
AS 12.29.25
Key words: Analytical photogrammetry

Considerations on the contents of the notion: photogrammetry; extension of this notion by considerations of the history of language with respect to iconometry or photogrammetry, under which we understand the evaluation of images; their physical law of representation may be modelled with respect to algebra. It leads to a generalization of the contents of photogrammetry. Finally, some ideas on the importance of computer-assisted measuring techniques for teaching and research are given.

Techn. Univ. Dresden IGD 84/III 83

528.85(202)

Serapinas, B. B.
Tochnost' komponentnogo analiza pri kosmicheskikh issledovanijakh prirodnojj sredy
Issl. Zemli Kosmosa. — Moskva (1982) 5. — p. 23—28
The accuracy of the component analysis in environmental studies from space (RUS)
AS 12.29.25
Key words: Environment, remote sensing

Dobrokhotov, National Correspondent, USSR IGD 84/III 84

528.235:515.591
528.711.7:629.783

Snyder, John P.
Space oblique Mercator projection mathematical development (ENG)
Geol. Survey Bull. — Washington (1981) 1518. — 108 p. : 1 lit.
Rzh 1982/5.52.100
AS 12.29.25/12.15.05
Key words: Automatic mapping, cartographical projection, remote sensing, scanning

The generation of the mathematical basis for mapping with extra-terrestrial photographs using scanner systems of Landsat type is dealt with. The results will be used in automated systems for mapping the Earth according to extra-terrestrial photographs.

Kombinat Geod./Kart. IWT, Leipzig IGD 84/III 85

528.726.6:528.721:513.012

Somogyi, J.
Az analitikus kiertekeloe mueszerekroel
Geod. es Kartogr. — Budapest 35 (1983) 3. — p. 160—163 : 1 fig. 2 tab. 3 lit.
On analytical plotters (HUN)
AS 12.29.25/73.11
Key words: Analytical plotter, orientation of photogrammes, photogrammetric plotter, stereo instrument

The paper gives a short review about the development of analytical plotters starting from Helava's invention. After a brief description of the basic principle, the function of the main parts of the system are discussed. Finally it gives a report on the existing types and on the capacity and accuracy of analytical instruments.

Veroe-Hetenyi, National Correspondent, Hungary IGD 84/III 86

528.8.042:550.8

Valjakh, V. M.
Aehrofotograficheskie i skanernye aehrometody pri inzhenerno-geologicheskikh issledovanijakh
Moskva: Nedra 1982. — 261 p. : 50 lit.
Aerophotographic and scanner aeromethods in engineering geological studies (RUS)
AS 12.29.25/12.29.41
Key words: Aerial photograph, geology, remote sensing, scanner

Dobrokhotov, National Correspondent, USSR IGD 84/III 87

528.711.7:528.236

Zobrist, Albert L. / Bryant, Nevin A. / McLeod, Ronald G.
Technology for large digital mosaics of Landsat data (ENG)
Photogramm. Engin. & Remote Sens. — Falls Church, Virginia 49 (1983) 9. — p. 1325—1335 : 7 fig. div. lit.
AS 12.29.25/12.29.29/12.15.05
Key words: Landsat digital data, map projection

Advances in algorithms and system executive procedures for digital image processing have made digital mosaicking of Landsat images (both MSS and RBV) an attractive possibility. The technology developed includes simultaneous map projection and adjustment of frame edges to eliminate both geometric and radiometric seams. The incorporation of ground control points, either by manual or automatic ground control point file identification, has resulted in root-mean-square (RMS) positional accuracies that exceed 1 : 100,000 scale National Map Accuracy Standards. Rotation to north vertical is accomplished at low computational cost. Input data frames can be cut in any arbitrary shape to remove cloud cover and accommodate terrain offset effects. Similarly, the final digital mosaic can be arbitrarily segmented to suit user requirements. Two large applications in Pennsylvania and Bolivia are reported. A test case utilizing RBV data is described.

(Author's summ.)

Techn. Univ. Dresden IGD 84/III 88

AIDOS Systematics **12.29.29** **Group III Photogrammetry and remote sensing**

Application of photogrammetric techniques to map production

Production and revision of topographic and thematic maps on all scales by means of aerial- and terrestrial-photogrammetric methods
Using of aerocosmic photographs to produce and revise maps

Gruppe III Photogrammetrie und Fernerkundung

Anwendung photogrammetrischer Verfahren zur Kartenherstellung

Herstellung und Laufendhaltung topographischer und thematischer Karten aller Maßstäbe mit aero- und terrestrisch-photogrammetrischen Verfahren
Nutzung aerokosmischer Aufnahmen zur Herstellung und Laufendhaltung von Karten

528.721.126:528.42(084.3)(988)
528.711.11:771.351.3:528.73

Bengtsson, T.
The mapping of northern Greenland (ENG)
Photogramm. Rec. – London 11 (1983) 62. – p. 135–150 : 13 fig. 11 lit.
AS 12.29.29
Key words: Aerial triangulation, orthophotomap, stereoplotter, super-wide angle photography

Approximately 110 sheets of orthophotomaps at a scale of 1 : 100,000 covering all the ice free area (300,000 km^2) of northern Greenland are to be produced by the Geodetic Institute of Denmark. Super-wide angle photography at a scale of 1 : 150,000 was acquired in 1978. Individual models are set up in a Wild B8S stereoplotter and, after carrying out aerial triangulation, a numerical method of absolute orientation is used. A digital ground model of the entire area is established in order to enable the interpolation of contour lines for automatic plotting and to produce control data for the orthophotoprojector, a Wild OR1. Methods of data handling and control, accuracies and typical errors are discussed.

(Author's summ.)

Techn. Univ. Dresden IGD 84/III 89

528.711:629.783
528.721.15

Fernandes, L.
Opyt ispol'zovanija kosmicheskikh snimkov dlja sozdanija fotoplanov
Geodez. i Kartogr. – Moskva (1982) 8. – p. 39–41 : 2 lit.
Experience with utilization of space photography for compilation of photoplans (RUS)
AS 12.29.29
Key words: Photoplan, space photography

Dobrokhotov, National Correspondent, USSR IGD 84/III 90

528.711(084.3):551.462

Fryer, J. G.
A simple system for photogrammetric mapping in shallow water (ENG)
Photogramm. Rec. — London 11 (1983) 62. — p. 203—208 : 2 fig. 14 lit.
AS 12.29.29
Key words: Depht co-ordinates, planimetric co-ordinates, shallow water, stereophotogrammetry
Techn. Univ. Dresden IGD 84/III 91

528.721.15:626.8

Petrakova, V. N. / Timushev, G. N.
Issledovanie variantov sozdanija fotoplana dlja proektirovanija meliorativnykh sistem
Geod., Kartogr. i Aehrofotos"jomka. — L'vov (1982) 36. — p. 125—129
Investigation of different versions of compilation of photomosaics for planning of meliorative systems (RUS)
AS 12.29.29/89
Key words: Melioration, photoplan
Dobrokhotov, National Correspondent, USSR IGD 84/III 92

528.74:528.4(084.3)

Rapasov, P. N.
Stereofotogrammetrija dlja celejj kartografirovanija i reshenija inzhenernykh zadach
Moskva: Nedra 1983. — 320 p. : 17 lit.
Stereophotogrammetry for map-making and engineering tasks (RUS)
AS 12.29.29/12.29.25
Key words: Engineering geodesy, stereophotogrammetry
Dobrokhotov, National Correspondent, USSR IGD 84/III 93

528.74:72

Trukhanov, A. E.
Primenenie dopolnitel'nykh uslovijj v processe fotogrammetricheskojj obrabotki snimkov inžhenernykh sooruzhenijj
Inzh.-Geod. raboty pri str-ve, montazhe i ehkspluat. inzh. sooruzh. — Moskva (1982). — p. 134—140
The use of additional constraints in the process of photogrammetric treatment of photographs of engineering constructions (RUS)
AS 12.29.29/83
Key words: Photogrammetric evaluation
Dobrokhotov, National Correspondent, USSR IGD 84/III 94

AIDOS Systematics **12.29.33 Group III Photogrammetry and remote sensing**

Application of photogrammetry for non-topographic purposes

Photogrammetric determination of the shape, size and position of non-topographic objects
Close-range photogrammetry, space-time photogrammetry
Application of unconventional imaging techniques in the close-range

Gruppe III Photogrammetrie und Fernerkundung

Anwendung der Photogrammetrie für nicht-topographische Zwecke

Photogrammetrische Form-, Größen- und Lagebestimmung nicht-topographischer Objekte
Nahphotogrammetrie, Raum-Zeit-Photogrammetrie
Anwendung nicht-konventioneller Abbildungsverfahren im Nahbereich

528.711(21-194.2).088

Abdel-Aziz, Youssef I.

The achieved accuracy from a closed-range photogrammetric system (ENG)

Techn. Papers 1982 ACSM-ASP Fall Convention, Hollywood, Florida, September 19—23, 1982. Ed.: American Congr. Surv. & Mapp. / American Soc. Photogramm. — Falls Church, Virginia 1982. — p. 1—15 : 2 fig. 7 lit.

AS 12.29.33

Key words: Close range photogrammetry

A photogrammetric system consists of many elements such as the camera, the plotter and the ground control points. The accuracy of the photogrammetric system is a function of the accuracies of these elements. The paper discusses the procedures for estimating the accuracies of these elements and the mathematical relation between the accuracies of these elements and the accuracy that will be achieved from the photogrammetric system.

(Author's summ.)

Techn. Univ. Dresden IGD 84/III 95

528.73(21-194.2):513.012

Earls, C. J.

Accuracy potential of a system for analytical close range photogrammetry (ENG)

Photogramm. Rec. — London 11 (1983) 62. — p. 169—182 : 6 fig. 3 tab. 13 lit.

AS 12.29.33/12.29.25

Key words: Bundle method, close range photogrammetry

This paper summarises the theoretical and practical development of an analytical system for close range photogrammetry. The system uses the extended bundle method of phototriangulation and a sophisticated least squares algorithm. A description is given of computer software, and aspects of statistical testing and gross error detection are discussed in order to assess the accuracy potential. Applications of the system for precise co-ordination of a target range and camera calibration are outlined.

(Author's summ.)

Techn. Univ. Dresden IGD 84/III 96

528.711(26.02)(261.2)
528.74:622.323(26.02)

Leatherdale, J. D. / Turner, D. J.

Underwater photogrammetry in the North Sea (ENG)

Photogramm. Rec. — London 11 (1983) 62. — p. 151—167 : 12 fig. 7 lit.

AS 12.29.33/12.29.25/12.29.37

Key words: Camera calibration, oil platform, underwater photogrammetry

Underwater photogrammetry is being used in the particularly hostile environment of the North Sea where it has been put into commercially viable operation. The authors discuss underwater photogrammetry related to other means of inspection of oil platforms. They also deal with operational experience of 35 mm and 70 mm cameras, camera calibration, the provision of control and photogrammetric observations and data analysis. The demand for underwater photogrammetry continues to grow.

(Author's summ.)

Techn. Univ. Dresden IGD 84/III 97

528.711.1:551.466

Mitchell, H. L.

Wave heights in the surf zone (ENG)

Photogramm. Rec. — London 11 (1983) 62. — p. 183—193 : 3 fig. 2 tab. 11 lit.

AS 12.29.33/12.29.09

Key words: Height measurement, ocean wave, terrestrial photogrammetry

The advantages and disadvantages of measuring ocean wave heights in the surf zone by using terrestrial photogrammetry are outlined. Difficulties arising from a lack of control points can be overcome by using the apparent horizon to provide camera orientations about two axes, provided that the apparent horizon is the sea surface and that an unconventional orientation of the camera axis system is adopted. Following practical tests and error simulation studies, height differences were determined for 12 points in the breaking wave zone off a Newcastle (New South Wales) beach. It is hoped that the method can be developed as a convenient means of collecting wave data.
(Author's summ.)
Techn. Univ. Dresden IGD 84/III 98

061.3(100+420)"1982"ISPRS:528.7/.8
528.7(21-194.2)

. . .
Precision and speed in close range photogrammetry (ENG)
ISPRS Sympos., York, England, 5th to 10th September, 1982
London: Internat. Archives Photogramm., 24 (1982) V/1 and 2. 557 p.
(Ed. J. W. C. Gates and K. B. Atkinson)
AS 12.29.33/12.01.17
Key words: Close range photogrammetry, ISPRS, symposium 1982, York England

The proceedings of symposium held by the Commission V of ISPRS in York, England 1982 are published as volumes of the International Archives of Photogrammetry. This volume contains 57 texts and papers submitted in the prescribed form by the due date. It is gratifying that so many authors seem to have drawn inspiration from the theme of "precision and speed in close range photogrammetry". Papers cover all six divisions of our current interest, although the largest group of papers relates to the Working Group on "Analytics". In terms of numbers also the papers relating to "Industrial Constructions and Mesuration" are not far behind.
Techn. Univ. Dresden IGD 84/III 99

528.7(21-194.2)

Wahl, M.
Photogrammetry at Regie Renault (ENG)
Presented at the Internat. Sympos. "Precision and Speed in Close Range Photogrammetry", Univ. of York, Sept. 1982
AS 12.29.33/12.29.25
Key words: Close range photogrammetry, Unisurf method

The Renault Company has developed a system for the mathematical definition of complex surfaces known as the Unisurf method. In association with this method, close range photogrammetry is used to provide the digital data required for the analysis of a surface. Photogrammetric techniques are of value in various fields of application, including the styling of new models of cars, design modifications to existing models, car accident investigations and statistical studies on human bodies.
(Author's summ.)
Techn. Univ. Dresden IGD 84/III 100

AIDOS Systematics **12.29.37** **Group III Photogrammetry and remote sensing**

Photointerpretation

Definition of characteristic features
Working stages of photointerpretation
Interpretation of air photographs and aerocosmic photographs
Application of photointerpretation
Automation of image interpretation
Application of technical aids to photointerpretation

Gruppe III Photogrammetrie und Fernerkundung

Photointerpretation

Begriffsbestimmung und Merkmale
Arbeitsstufen der Photointerpretation
Interpretation von Luftbildern und photographischen aerokosmischen Aufnahmen
Anwendung der Photointerpretation
Automatisierung der Bildinterpretation
Anwendung technischer Hilfsmittel in der Photointerpretation

528.77.063.9:55
528.711:629.783

Beloborodov, M. A. et al.
Metody matematicheskojj obrabotki materialov deshifrirovanija kosmicheskikh snimkov i geologo-geofizicheskojj informacii pri prognoze ehndogennogo orudenenija
Genezis rud. mestorozhd. Tez. dokl. 6 Simpoz. MAGRM, Tbilisi 1982. — Tbilisi (1982). — p. 167—168
The methods of mathematical processing of space imagery interpretation data and geological and geophysical information for prediction of endogenetic ore formation (RUS)
AS 12.29.37/51.19
Key words: Aerial photography interpretation, geology, geophysics, satellite
Dobrokhotov, National Correspondent, USSR IGD 84/III 101

528.711.7:622.34

Carr, James R. / Glass, Charles E. / Schowengerdt, Robert A.
Signature extension versus retraining for multispectral classification of surface mines in arid regions (ENG)
Photogramm. Engin. & Remote Sens. — Falls Church, Virg. 49 (1983)
AS 12.29.37/12.29.41/12.23.25
Key words: Inventory of mined land, land classification, Landsat, spectral signature

Because governmental agencies are responsible for conducting mined land inventories on a regular basis, it is desirable to standardize the manipulation of the digital Landsat data for this purpose. Supervised classification of Landsat images was used in this study to map Arizona copper mining activity in order to demonstrate the utility of these images for dynamic regional inventories of mined lands. A set of spectral signatures derived at a single large mining complex were extended spatially and temporally to discriminate mining activity from natural terrain features at other sites. Accurate classifications were obtained using signature extension, and in some cases, the accuracy exceeded that obtained by retraining the classifier for individual mines.

(Author's summ.)

Techn. Univ. Dresden IGD 84/III 102

528.77:622.323(202)

Lastochkin, A. N.
Deshifrirovanie po kosmosnimkam glubinnykh platformennykh struktur, perspektivnykh na neft' i gaz
Sov. geod. (1982) 12. — p. 116—121
Interpretation of space imagery for detection of oil- and gasbearing deep block structures (RUS)
AS 12.29.37
Key words: Photo interpretation
Dobrokhotov, National Correspondent, USSR IGD 84/III 103

528.711.7:63(73)
528.8:629.783

Loveland, Thomas R. / Johnson, Gary E.
The role of remotely sensed and other spatial data for predictive modeling: the Umatilla, Oregon, example* (ENG)
Photogramm. Engin. & Remote Sens. — Falls Church, Virgin. 49 (1983) 8. — p. 1183—1192 : 2 fig. 7 tab. 5 plate. 9 lit.
AS 12.29.37/12.29.25
Key words: Aerial photography, irrigation agriculture, Landsat

Landsat data and 1 : 24,000-scale aerial photographs were initially used to map the expansion of irrigation from 1973 to 1979 and to identify crops under irrigation in 1979. The crop data were then used with historical water requirement figures and digital topographic and hydrographic data to estimate water and power use for the 1979 irrigation season. The final project task involved production of a composite map of land suitability for irrigation development based on land cover (from Landsat), landownership, soil irrigability, slope gradient, and potential energy costs.

The methods and data used in the study demonstrated the flexibility of remotely sensed and other spatial data as input for predictive models. When combined, they provided useful answers to complex questions facing resource managers.

* Presented at the Pecora VII symposium, Sioux Falls, South Dakota, 18—21 October 1981.

(Author's summ., abbr.)

Techn. Univ. Dresden IGD 84/III 104

528.711.7:634

Nelson, Ross F.
Detecting forest canopy change due to insect activity using Landsat MSS (ENG)
Photogramm. Engin. & Remote Sens. — Falls Church, Virginia 49 (1983) 9. — p. 1303—1314 : 3 fig. 4 tab. div. lit.
AS 12.29.37/12.29.41/87
Key words: Forestry, Landsat

Multitemporal Landsat multispectral scanner data were analyzed to test various computer-aided analysis techniques for detecting significant forest canopy alteration. Three data transformation — differencing, ratioing, and a vegetative index difference — were tested to determine which best delineated gypsy moth defoliation. Response surface analyses were conducted to determine optimal threshold levels for the individual transformed bands and band combinations. Results indicate that, of the three transformations investigated a vegetative index difference (VID) transformation most accurately delineates forest canopy change. Band 5 (0.6 to 0.7 μm) ratioed data did nearly as well. However, other single bands and band combinations did not improve upon the band 5 ratio and VID results.

(Author's summ.)

Techn. Univ. Dresden IGD 84/III 105

AIDOS Systematics **12.29.41** **Group III Photogrammetry and remote sensing**

Non-photographic recording of information from the air and its interpretation

Recording of information by non-photographic techniques and aerocosmic methods
Using of recording and imaging devices
Interpretation of non photographic images
Decoding of photoelectronic images
Application of the results of interpretation

Gruppe III Photogrammetrie und Fernerkundung

Nichtphotographische Erfassung von Informationen aus der Luft und ihre Interpretation

Erfassung von Informationen mit nichtphotographischen Verfahren und aerokosmischen Methoden
Einsatz von Erfassungs- und Aufnahmegeräten
Interpretation nichtphotographischer Aufnahmen
Dechiffrierung photoelektronischer Aufnahmen
Anwendung der Interpretationsergebnisse

528.711.7:634(73)

Blanchard, W. Anthony
Identification and mapping of riparian woodlands from simulated thematic mapper data (ENG)
Techn. Papers 1982 ACSM-ASP Fall Convention, Hollywood, Florida, September 19—23, 1982. Ed.: American. Congr. Surv. & Mapp. / American Soc. Photogramm. — Falls Church, Virginia 1982. — p. 50—61 : 14 fig. div. lit.
AS 12.29.41/12.29.37
Key words: Forestry, Landsat

Simulated Landsat-D Thematic Mapper ditigal data provided by the NASA Earth Resources Laboratory (ERL) is used to identify and map riparian woodland habitats in north central Oklahoma. Three digital image enhancement techniques are applied to the Thematic Mapper data to increase the ability to descriminate riparian vegetation from other vegetation communities. A linear contrast stretch is used to increase the dynamic range of the channel 3 (0.63—0.69 μm) data. A haze correction algorithm is applied to the channel 2 (0.53—0.60 μm) data, and band ratios of other channels are attempt to enhance identification accuracies.
(Author's summ., abbr.)

Techn. Univ. Dresden IGD 84/III 106

528.711.7(26):621.396.96

Kasischke, Eric S. et al.
Detection of bottom features on Seasat synthetic aperture radar imagery (ENG)
AS 12.29.41/12.29.25/12.15.21
Key words: Seasat synthetic operture radar

A distinct set of surface patterns detected in imagery collected by the Seasat synthetic aperture radar (SAR) have been shown to be the result of an interaction between a physical oceanic process and a bottom topographic feature. These patterns can be used to infer a bottom feature, and thus are a potential source of information for identifying submerged features hazardous to surface navigation. The Seasat SAR imaged bottom-related surface patterns in both deep and shallow water. Examples of these surface patterns are presented in this paper along with explanations of how they occur.
(Author's summ.)

Techn. Univ. Dresden IGD 84/III 107

528.854.2

Schulz, Bernd-Siegfried
Noise reduction and line following in multi-scanned data (ENG)
Paper presented at Internat. Sympos., Helsinki, June 7—11. 1982. Commission III
Nachr. Kt.- u. Vermess.-Wesen, R. II. — Frankfurt/M. (1983) 40. — p. 109—118 : 2 fig. 1 tab. 1 lit.
AS 12.29.41/12.29.25
Key words: Noise reduction, remote sensing, scanner

The author presents a method not specific to application which serves, during the pre-processing, for clearing digital recordings to a large extent from the noise caused by the electrooptic scanning as well as by the object-dependent aerial inhomogeneity of the grey values.

Hereby the logical connection of functions of the initial data works against the two-dimensional incidental distribution of the electronic noise, whereas the averaging of the grey values within threshold-dependent intervals existing between successive significant deficiencies works against the object noise.

(Author's summ., abbr.)

Techn. Univ. Dresden IGD 84/III 108

Bei Fragen zur Produktsicherheit wenden Sie sich bitte an:
If you have any questions regarding product safety,
please contact:

Walter de Gruyter GmbH
Genthiner Straße 13
10785 Berlin
productsafety@degruyterbrill.com